MARSOC

U.S. Marine Corps Special Operations Command

FRED PUSHIES

ZENITH PRESS

To the Silent Warriors... Semper Fi!

First published in 2011 by Zenith Press, an imprint of MBI Publishing Company, 400 First Avenue North, Suite 300, Minneapolis, MN 55401 USA

© 2011 Zenith Press
Text © 2011 Fred Pushies

All rights reserved. With the exception of quoting brief passages for the purposes of review, no part of this publication may be reproduced without prior written permission from the Publisher. The information in this book is true and complete to the best of our knowledge.

Zenith Press titles are also available at discounts in bulk quantity for industrial or sales-promotional use. For details write to Special Sales Manager at MBI Publishing Company, 400 First Avenue North, Suite 300, Minneapolis, MN 55401 USA.

To find out more about our books, join us online at www.zenithpress.com.

Library of Congress Cataloging-in-Publication Data

Pushies, Fred J., 1952–
  MARSOC : U.S. Marine Corps Special Operations Command / Fred Pushies.
    p. cm.
  Includes index.
  ISBN 978-0-7603-4074-5 (pbk.)
 1. United States. Marine Special Operations Command. 2. Special forces (Military science)—United States. I. Title.
  VE23.P92 2011
  359.9'6--dc22
                         2011016255

About the author: Fred Pushies has spent two decades in the company of units from Navy SEALs to Army Night Stalkers to Force Recon Marines. His previous books include *10th Mountain Division*, *82nd Airborne*, *Marine Force Recon*, *Special Ops*, and *Weapons of Delta Force*. He lives outside Detroit.

All photographs are from the author's collection unless noted otherwise.

On the cover: MARSOC Marines come ashore under the cover of fog. *USMC photo, MARSOC PAO*

On the frontispiece: Marines shoot M-4 carbines as part of a five-week marksmanship and close quarters battle course, part of the seven month-long MARSOC Individual Training Course. *USMC photo, Sgt. Adam Sanchez*

On the title page: A MARSOC Marine takes cover as a sand storm approaches in Farah province, Afghanistan, February 2010. *USAF photo, Staff Sgt. Nicholas Pilch*

On the back cover:
Main: A Marine from Special Operations Task Force West investigates a possible threat while on patrol in the village of Zanghlav, Afghanistan. *USMC photo, Sgt. Brian Kester*

Inset: Marines from Special Operations Task Force West on patrol near Zanghlav and Rabat-I-Sapcha villages, Afghanistan. *USMC photo, Sgt. Brian Kester*

Printed in China

CONTENTS

chapter 1 ORIGIN OF SOCOM page 6

chapter 2 HISTORY OF MARINE CORPS SPECIAL OPERATIONS page 22

chapter 3 MARSOC CREATION AND ORGANIZATION page 42

chapter 4 SELECTION AND TRAINING page 62

chapter 5 WEAPONS page 86

chapter 6 TACTICAL GEAR page 122

chapter 7 TACTICS, TECHNIQUES AND PROCEDURES page 142

chapter 8 MARSOC INTO THE FUTURE page 150

ACKNOWLEDGMENTS page 155

ABBREVIATIONS page 156

TERMINOLOGY ("MARINE SPEAK") page 158

INDEX page 160

CHAPTER 1
ORIGIN OF SOCOM

IN THE EARLY MORNING HOURS of May 2, 2011 (Pakistan time), an elite team of American commandoes from SEAL Team Six completed a special operations mission that killed Osama bin Laden, the head of al Qaeda, America's most implacable enemy in the global war on terror. Superb intelligence, training, preparation, coordination, and execution are the hallmarks of this success. Although this is only the most notable of America's special operations victories, it serves as confirmation that the future of modern warfare will be increasingly in the hands of our special operations forces, "the quiet professionals."

It was not always so.

In November 1979, a group of radical Iranian "students" captured the U.S. embassy in Tehran, Iran. The United States Air Force rushed to regenerate its special operations capabilities. By December 1979, a rescue force was chosen, and training commenced. Training exercises were conducted through March 1980, and on April 16, 1980, the Joint Chiefs of Staff (JCS) approved the mission. On the nineteenth, rescue forces consisting of Army, Navy, Air Force, and Marine assets began to deploy to Southwest Asia. Delta Force, as this composite unit was called, was tasked with the assault on the embassy and the rescue of the American hostages.

On April 24, after six months of failed negotiations, the National Command Authority (NCA) ordered the execution of Operation Eagle Claw to free U.S. hostages held in Iran. Under cover of darkness, eight RH-53D helicopters departed the aircraft carrier USS *Nimitz* on station in the Arabian Sea; at the same time, six C-130s left Masirah Island, Oman. Both sets of aircraft set off for a prearranged site six hundred miles into the desert, code designation "Desert One." The Achilles' heel in the operation proved to be the helicopters; a few hours into the mission, two of them had aborted due to mechanical failure. A desert dust storm, known as a *haboob*, caused the remaining helicopters to arrive late, and yet another suffered a hydraulic leak that its crew determined was unfixable at the Desert One site. This determination resulted in only five operational helicopters, and mission planners had ascertained a minimum of six were required for the mission to continue. Now, with only five available, the mission was aborted.

Many books and articles have been written describing the mission in detail, so we will not belabor the operation here. Simply stated, the plan was to assemble eight Navy RH-53D helicopters at a predetermined location in the Iranian wasteland code-named "Desert One." During the night, the helicopters would be refueled from KC-130 tankers (also having landed in the desert), load a 120-man Army assault team aboard, and proceed to two additional hide sites. The Delta assault team would continue on to the U.S. embassy, extract the hostages, rendezvous with the helicopters, and be extracted out of the city. That never happened.

With the decision to abort, it was time to load up the aircraft and "get out of Dodge." While repositioning his RH-53D helicopter for refueling, one of the pilots collided with a C-130. This resulted in the two aircraft rapidly being engulfed in flames. With the situation hastily moving from bad to worse to disastrous, the on-scene commander, Col. Charlie Beckwith, decided to load all the survivors,

OPPOSITE: Three RH-53 Sea Stallion helicopters line up on the flight deck of the nuclear-powered aircraft carrier USS *Nimitz* (CVN-68) in preparation for Operation Eagle Claw. The covert mission's objective was to rescue American hostages being held in the U.S. embassy in Tehran, Iran. DOD PHOTO

remaining troops, assault team, and Marine aircrews onto the C-130 and depart Desert One as soon as possible. Eight men had been killed and five more injured. Five intact helicopters, the burned wreckage of the helicopter and C-130, and the dead soldiers were left behind. Operation Eagle Claw had failed and cost the lives of eight gallant troops. It would also cost the honor of the United States of America and the credibility of U.S. Special Operations.

The disaster that transpired at Desert One was the culmination of years of post–Vietnam era thinking among U.S. commanders that had led to the decline of special operations capabilities during the 1970s. During that time, there was a marked distrust between special operations forces and the conventional military. This was further denigrated by significant cuts in funding for special operations forces (SOF) units and missions.

Following the disaster at Desert One, the Department of Defense appointed a review committee known as the Holloway Commission to look into problems within U.S. Special Operations. The commission was chaired by Adm. James L. Holloway III and included Gen. Leroy Manor, who had commanded the Son Tay raid into North Vietnam. The outcome of this inquiry resulted in two major recommendations. First, the Department of Defense needed to establish a counterterrorism joint task force (CTJTF) as a field organization of the JCS with a permanently assigned staff and forces. The JCS would plan, train for, and conduct operations to counter terrorist activities directed against the United States. The CTJTF would utilize military forces ranging in size from small units of highly specialized personnel to larger integrated forces. Second, the JCS needed to consider the formation of a Special Operations Advisory Panel (SOAP) consisting of high-ranking officers to be drawn from both active service and retired personnel. The prerequisite for selection was a background in special operations or service at a commander in chief (CINC) or JCS level, and participants needed to maintain a proficient level of interest in special operations or defense policy.

One issue raised by the commission concerned the selection of helicopter aircrews. Were the pilots up to the task? Why had they selected the Marines when more than a hundred qualified Air Force H-53 pilots were available? If we look at today's standards, the Marine pilots had big boots to fill. They were being asked to fly at night, which was unusual practice for the "flying leathernecks." These pilots were asked to launch off the deck of a carrier *at night*, fly nap-of-the-earth where radar could not detect them, and fly without running lights. The pilots were issued PVS-5 night vision goggles; however, they could only be worn at thirty-minute intervals. This meant the pilot and copilot had to alternate flying the huge helicopter every thirty minutes. The Marines had no pilots that had been trained in this type of flying. In fact, none of the services were prepared for such a contingency. Yet the Marines did what they always do: they adapted, overcame, and improvised.

Regarding the helicopters, the commission concluded that a minimum of ten to twelve helicopters should have been deployed. This redundancy in air assets would have secured a higher probability of success if problems arose, assuring that six helicopters, the minimum requirement for completion of the mission, would be operational at Desert One. The legacy of this decision may prove to hamper future special operations. It was this commission's direction that spawned interest in the V-22 Osprey, redesignated the MV-22 and CV-22 for use in Marine and Special Operations Forces.

Another issue was the lack of a thorough readiness assessment and schedule of mission rehearsals. From the onset, training was not conducted in a truly joint method. Due to security and logistical considerations, the training was compartmentalized and held at scattered locations across the continental United States, as well as abroad. The limited rehearsals conducted could only evaluate various segments of the entire mission. Additionally, preparation was carried out at the individual and unit levels within each element.

There was no designated mission commander for six months, breaching the principle of unity of command. This lack of command and control hampered the training, planning, and execution of the operation. There were separate commanders for site security, ground force, landing support, KC-130s, and the helicopter force. Compounding the problem were procedural limitations and communications interoperability.

US SOCOM organization chart. MARSOC

While the mission itself ended in disaster, Desert One did serve to strengthen the resolve of some within the Department of Defense to reform SOF. As Lt. Gen. James Vaught (Ret.), who commanded the rescue task force, commented in later years: "Eagle Claw was a successful failure. We wanted with all our being to rescue the Americans. However, had we succeeded, conventionalists in all likelihood would have said we did not need a full-time training and ready force which could quickly and successfully rescue Americans the world around. [Without Desert One] we would not have the competent, proven, ready special operations forces that are today the envy of the world."

Indeed, Desert One served as a catalyst for the evolution of the U.S. Special Operations Forces, and seven years later, on April 13, 1987, President Ronald Reagan approved the establishment of the new command. The Department of Defense activated USSOCOM on April 16, 1987, and nominated Gen. James J. Lindsay to be the first Commander in Chief, Special Operations Command (USCINCSOC).

USSOCOM

Overseeing all of the U.S. Special Operations Forces is the U.S. Special Operations Command (USSOCOM), located at MacDill Air Force Base in Tampa, Florida. Today, each of the services—Army, Air Force, Navy, and Marines—has a command subordinate to SOCOM. An additional subordinate command under SOCOM is the Joint Special Operations Command (JSOC).

The mission of SOCOM is to provide capable special operations forces (SOF) to defend the United States and its interests. SOCOM also plans and synchronizes Department of Defense (DOD) operations against terrorist networks. America's SOF units are organized, equipped, trained, and then deployed by SOCOM to meet the high operational demands of geographic combatant commanders (GCCs) around the globe. To accomplish those missions, SOCOM focuses on three priorities: The first is to deter, disrupt, and defeat terrorist threats to the United States. This is accomplished by planning and conducting special operations, emphasizing culturally attuned international engagement, and fostering interagency cooperation. Second, the command is to develop and support the SOF members as well as their families. Third, SOCOM must sustain and modernize the SOF units by equipping the operators, upgrading their mobility platforms, and further developing continuing intelligence, surveillance, and reconnaissance (ISR) sensors and systems. These priorities support SOCOM's ongoing efforts to ensure SOF members are highly trained, properly equipped, and deployed to the right places at the right times for the right missions.

SOCOM units are capable of planning and conducting a variety of lethal and nonlethal special operations missions, both complex and ambiguous in austere environments. The U.S. SOF is made up of small units that work alone or in combination with one another in both direct and indirect military operations, often utilizing tactics of unconventional warfare. These "quiet professionals," as they are often called, are trained in the newest methods, tactics, and procedures, and are equipped with the latest technology and weaponry. The soldiers, sailors, Marines, and airmen who constitute SOCOM units exceed the capabilities of conventional military forces. Each of the services' units cross-train in many of the same techniques, tactics, and procedures (TTP), and there are times when their missions overlap. Their missions are frequently clandestine in nature and often politically sensitive. In military operations other than war (MOOTW), it may be necessary to deploy a small force to stealthily accomplish missions directed from National Command Authority. Such missions would be assigned to the appropriate SOF unit.

UNITED STATES ARMY SPECIAL OPERATIONS COMMAND (USASOC)

Special Forces

U.S. Army Special Forces Command (Airborne) (USASFC[A]) is comprised of seven major subordinate units. Each of these units, known as Special Forces groups,

7th Special Forces Group (Airborne) soldiers move alongside each other during urban combat training at Fort Bragg, North Carolina. They are armed with the MK 18 carbine, a modified M4A1 with a shortened 10.3-inch barrel and fitted with a close quarters battle receiver (CQBR). U.S. ARMY PHOTO, SGT. DANIEL LOVE, 7TH SFG(A) PAO

Soldiers from the 75th Ranger Regiment (Airborne) descend in an MH-6 Little Bird helicopter flown by pilots from the 160th Special Operations Aviation Regiment (Airborne) during an exercise demonstrating the range of U.S. Army Special Operations capabilities at Fort Bragg, North Carolina. USASOC PHOTO, TRISH HARRIS, PAO

is commanded by a colonel. The mission statement of the SF Command is "to organize, equip, train, validate and prepare forces for deployment to conduct worldwide special operations, across the range of military operations, in support of regional combatant commanders, American ambassadors and other agencies as directed."

In addition to the military skills the SF soldiers possess, they are schooled in the cultures, traditions, and languages of the regions in which they operate. This ability makes them aptly proficient in behind-the-lines, covert, and unconventional warfare missions.

Rangers

The 75th Ranger Regiment is the Army's premiere light infantry rapid assault force assigned to SOCOM. The unit's primary mission revolves around DA (or direct action) predominant operations. The soldiers of the 75th have one purpose in life: to kill the enemy and break things. Their specialty is airfield seizure, though they are also more than capable of conducting raids, combat search and rescue (CSAR), special equipment, and performing other light infantry operations. They may be inserted and extracted by land, sea, and air. Ranger units focus on

A U.S. Navy SEAL, armed with the FN special operations assault rifle–heavy (SCAR-H), which is chambered for 7.62mm ammunition, takes up a defensive position in a village in northern Zabul province, Afghanistan, on April 10, 2010. Afghan National Army soldiers, assisted by U.S. Special Operations members, investigated the presence of drug facilities in the province. U.S. NAVY PHOTO, MCCS JEREMY L. WOOD

mission-essential tasks, including movement to contact, ambush, reconnaissance, airborne and air assaults, and hasty defense.

The 75th Regiment maintains a constant state of readiness, and can be deployed anywhere in the world. On any given day, one Ranger battalion serves as Ready Reaction Force (RRF) 1 with the requirement to be "wheels up" within eighteen hours of notification. Additionally, one rifle company with battalion command and control can deploy in nine hours.

The 160th Special Operations Aviation Regiment (Airborne)

The 160th Special Operations Aviation Regiment (Airborne) (SOAR[A]) provides aviation support to U.S. Special Operations Forces. Primarily an Army asset, the 160th has a close working relationship with other units under the SOCOM command. The majority of its missions are carried out under the cover of darkness; hence, the unit has earned the nickname of the "Night Stalkers." The 160th Regiment's equipment includes MH-6 Little Bird light assault helicopters, MH-60 Blackhawk utility helicopters, and MH-47 Chinook medium-lift helicopters. Additionally, some of the regiment's Little Bird and Blackhawk helicopters, respectively known as AH-6 and MH-60L DAP, have been configured for close air support (CAS) missions. The unit's specialty is the covert insertion, resupply, and extraction of SOF teams. Additionally, the soldiers may conduct armed escort, reconnaissance, surveillance, and electronic warfare in support of missions.

The regiment is home to the "hottest" aviators in the Army, and, some operators will go on to add, "in the world."

The Night Stalkers maintain several types of helicopters in their inventory, from the small and agile Little Birds to the larger Chinooks, so that if SOF needs to perform a fast rope insertion/extraction system (FRIES) on a rooftop, request CAS, or extract from a "hot" landing zone, the 160th SOAR(A) has the aircraft, pilots, and aircrews to accomplish the task, plus or minus thirty seconds. The unit's motto is "Night Stalkers don't quit!" These consummate flyers are the perfect complement to the covert warriors of SOF.

NAVAL SPECIAL WARFARE COMMAND

Navy SEALs

The term *SEAL* is an acronym for **SE**a, **A**ir, and **L**and. Navy SEALs are qualified in diving and parachuting and are experts at combat swimming, navigation, demolitions, weapons, and many other skills. SEALs operate in small units called platoons. Squad size is typically eight men with two squads per platoon. In addition to the maritime environment, SEALs train in the desert, in the jungle, in cold weather, and in urban surroundings. Their forte is the

Two special operations craft riverine (SOC-R) boats cruise along the Salt River during live fire training. Naval special warfare combatant-craft crewmen (SWCC) from Special Boat Team 22 (SBT-22) employ the SOC-R, which is specifically designed for the clandestine insertion and extraction of Navy SEALs and other Special Operations Forces along shallow waterways and open water environments. U.S. NAVY PHOTO, MCCS ROBYN B. GERSTENSLAGER

quick DA raid, and they usually are not equipped for long, protracted contact with the enemy.

Special Boat Teams

Special warfare combatant crewmen (SWCC) are members of special boat teams (SBT), which pilot a variety of special warfare craft, from brown-water riverine vessels to the blue-water Mark V special operations craft (SOC). While they aren't SEALs, they do support SEAL personnel and other SOF units during their maritime and riverine missions, and they conduct unconventional small boat operations, such as coastal and riverine patrols.

A selected number of SEALs are tasked with the operation of the SDV, or SEAL delivery vehicle. These are small submersible craft that allow the SEALs to covertly infiltrate an enemy harbor or other waterborne target.

AIR FORCE SPECIAL OPERATIONS COMMAND

AFSOC is the Air Force element of SOCOM, and its mission is to provide mobility, surgical firepower, covert tanker support, and special tactics teams. These units will normally operate in concert with U.S. Army and Navy Special Operations Forces, including Special Forces, Rangers, the Special Operations Aviation Regiment, SEAL teams, PSYOP forces, and civil affairs units. AFSOC supports a wide range of activities from combat operations of a limited duration to longer-term conflicts. The unit also provides support to foreign governments and their militaries. Dependent on shifting priorities, AFSOC maintains a flexible profile, allowing it to respond to numerous missions.

AFSOC brings the "big guns" in as close air support of any SOF mission. With the AC-130H Spectre and the flagship of the 4th Special Operations Squadron, the AC-130U Spooky gunships, bristling with 25mm chain gun, 40mm Bofors cannon, and 105mm howitzer, the unit brings new meaning to the phrase, "Death from above." An AFSOC combat controller on the ground and a gunship overhead are among the most lethal weapons in the SOCOM arsenal.

Special Tactics Squadron

In addition to its awesome airpower, AFSOC may deploy special tactics squadrons composed of combat controllers and pararescuemen. Special tactics teams (STT) are proficient in sea-air-land insertion tactics into forward, nonpermissive environments. Combat control teams (CCT) establish assault zones with an air traffic control (ATC) capability. An assault zone can be a drop zone (DZ) for a parachute deployment or an LZ for heliborne

An AC-130U gunship of the 4th Special Operations Squadron. The gunships of the Air Force Special Operations Command provide close air support (CAS) for Special Operations Forces. Their ability to loiter in an area for an extended period of time makes them ideal weapons platforms in support of SOF missions. USAF PHOTO

operations or follow-on fixed-wing aircraft. It can also be for an extraction or low-level resupply.

The combat control teams are also responsible for ground-based fire control of the AC-130 Spectre gunships and helicopters of AFSOC, as well as having responsibility for *all* air assets, including Army and Navy aircraft.

In addition to these capabilities, CCTs provide vital command and control capabilities in the forward area of operations (AO) and are qualified in demolition to remove obstructions and obstacles to the LZ/DZ.

Also included under the command are the special operations weather teams (SOWT). Their mission is to provide meteorological and oceanographic information in and for the special operations theater of operations. Their functions include tactical infiltration, data collection, analysis and forecasting, mission tailoring of environmental data, and joint operation with host nation weather personnel. SOWT personnel perform this job from forward-deployed bases or, at times, from behind enemy lines using tactical weather equipment and an assortment of communications equipment.

JOINT SPECIAL OPERATIONS COMMAND (JSOC)

JSOC was created in 1980 and "officially" is a joint headquarters designated specifically to analyze U.S. Special Operations requirements and techniques, as well as to ensure interoperability and equipment standardization for U.S. SOF units. JSOC plans and conducts joint special operations exercises and training; the command also develops joint special operations tactics, equipment, and development. Also operating under JSOC are two units so secret that the Department of Defense does not even acknowledge their existence. These are the Army's Delta Force and the Navy's Development Group. Although each unit has an expansive capability, and latitude, their primary mission is counterterrorism (CT).

U.S. MARINES AND SOCOM

When SOCOM was created in 1987, it became responsible for all special operations forces in the U.S. military (i.e., Army Special Forces, Rangers, and the 160th Special Operations Aviation Regiment; Air Force Special Operations wings and squadrons; and Navy SEALs). Although the Navy placed the SEALs under this new command, the Marines would not be included.

An interesting fact is that prior to the official creation of SOCOM, a member of the U.S. Congress approached

Air commandos from the 720th Special Tactics Squadron perform fast-rope insertion methods from a CV-22 Osprey tilt-rotor aircraft from the 8th Special Operations Squadron. Both units are based at Hurlburt Field, Florida. USAF PHOTO

Gen. Paul X. Kelley with the proposition that all SOF units be placed under the command of the Marine Corps. The commandant was not receptive to the proposal, explaining that it would impact the Marine Corps mission around the world. Furthermore, while the Marines would not become plank owners of SOCOM, Force Reconnaissance Marines were trained in the same special operations skills as those in the SOF community.

There are a number of theories why the Marines were not included. One is that SOCOM had all aspects of the battlefield covered—land (Army), sea (Navy), and air (Air Force)—hence, the addition of the Marines would have been a duplication of forces. The most popular theory relates that General Kelley, who was the commandant at the creation of SOCOM, did not want to relinquish command and control of the Marines to an Army general. It is also a known fact that during the reorganization, many members of Congress, who themselves had served in the Corps, preferred the Marines to maintain their autonomy. While the Force Recon Marines processed the same special operations capabilities as their sister services, the consensus was that eventually Force Recon would end up being absorbed into SOCOM and would no longer be available to support Marine Expeditionary Units (MEUs) or Marine Expeditionary Forces (MEFs). Consequently, the Marines did not become part of SOCOM and were not categorized as special operations forces but rather were special operations capable, or SOC; hence, the designator MEU (SOC).

In January 2002, the commandant presented the first official plans outlining the permanent force contribution the Marines could bring to SOCOM. In the course of developing this concept, three potential plans of action materialized: First, the Marines would support SOCOM with the unique capabilities of their forward-deployed Marine Air-Ground Task Forces (MAGTFs). These MAGTFs would augment and sustain SOF capabilities. Second, Marine units would be OPCON (operational control) to SOCOM on a recurring or rotational basis to execute designated missions. Third, the Corps would provide Marine forces to be permanently assigned to SOCOM as part of a Marine Special Operations component, or MARSOC.

General Paul Xavier Kelley served as the commandant of the Marine Corps from July 1983 until June 1987. Word on the street, or rather in the foxholes, was that General Kelley did not want to relinquish command of his Marines to a separate command under the control of a U.S. Army general. Hence, the Marines were not included when SOCOM was created. USMC PHOTO

On February 24, 2006, the Marine Corps Forces Special Operations Command (MARSOC) officially became the Marine component of SOCOM. MARSOC conducts the manning, organizing, training, and equipping of Marine Special Operations Forces to accomplish its mission. The MARSOC headquarters, located at Camp Lejeune, North Carolina, is responsible for identifying Marine special operations–unique requirements; developing MARSOC tactics, techniques, procedures, and doctrine; and executing assigned missions in accordance with designated conditions and standards.

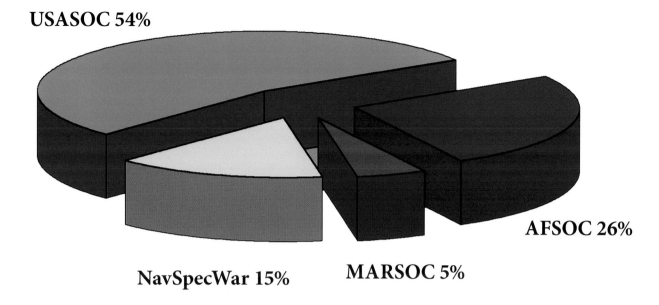

Percentage chart of SOCOM service forces. MARSOC

SPECIAL OPERATIONS PRINCIPAL MISSIONS

According to the SOCOM posture statement, nine activities have been designated as special operations principal missions: direct action (DA), combating terrorism (CBT), foreign internal defense (FID), unconventional warfare (UW), special reconnaissance (SR), information operations (IO), civil affairs (CA), counterproliferation of weapons of mass destruction (CP), and psychological operations (PSYOP). SOF is organized, trained, and equipped specifically to accomplish these nine tasks. These tasks as related by SOCOM are listed below.

Direct Action (DA)

DA operations include short-duration strikes and other small-scale offensive operations principally undertaken by SOF to seize, destroy, capture, recover, or inflict damage on designated personnel or matériel. While conducting these operations, SOF may employ raid, ambush, or direct assault tactics; emplace mines and other munitions; conduct stand-off attacks by fire from air, ground, or maritime platforms; provide terminal guidance for precision weapons; and conduct independent sabotage and antiship operations.

Combating Terrorism (CBT)

CBT is a highly specialized, resource-intensive mission. Certain SOF units maintain a high state of readiness to conduct CBT operations and possess a full range of CBT capabilities. CBT activities include antiterrorism (AT), counterterrorism (CT), recovery of hostages or sensitive material from terrorist organizations, attacks on terrorist infrastructure, and reduction of vulnerability to terrorism.

Foreign Internal Defense (FID)

FID involves participation by civilian and military agencies of a government in any of the action programs taken by another government to free and protect societies from subversion, lawlessness, and insurgency. SOF's primary contribution in this interagency activity is to organize,

A member of the Marine Special Operations Battalion (MSOB) trains for a direct action (DA) raid. The Marine Special Operations Command (MARSOC) officially became an active component of SOCOM in 2006.

HOW MARSOC FITS SOF CORE VALUES

Quality is better than quantity
MARSOC will formalize training procedures to ensure MARSOC Marines meet or exceed SOF standards.

Competent SOF cannot be created after emergencies occur
MARSOC will take decisive, deliberate methodical steps to increase U.S. special operations capabilities so its Marines and sailors are better able to anticipate and prevent emergencies in the future.

Humans are more important than hardware
MARSOC will establish clear screening, assessment, and selection standards to ensure MARSOC Marines are mature, experienced, capable warfighters.

SOF cannot be mass-produced
MARSOC will build its capabilities deliberately and methodically.

train, advise, and assist host nation military and paramilitary forces. The generic capabilities required for FID include instructional skills; foreign language proficiency; area and cultural orientation; tactical skills; advanced medical skills; rudimentary construction and engineering skills; familiarity with a wide variety of demolitions, weapons, weapon systems, and communications equipment; and basic PSYOP and CA skills.

Unconventional Warfare (UW)

UW includes guerrilla warfare, subversion, sabotage, intelligence activities, evasion and escape, and other activities of a low visibility, covert, or clandestine nature. When UW is conducted independently during conflict or war, its primary focus is on political and psychological objectives. When UW operations support conventional military operations, the focus shifts to primarily military objectives.

Special Reconnaissance (SR)

SOF conducts a wide variety of information-gathering activities of strategic or operational significance. Collectively, these activities are called SR. SR complements national and theater intelligence collection systems by obtaining specific, well-defined, and time-sensitive information when other systems are constrained by weather, terrain-masking, hostile countermeasures, or conflicting priorities. SR tasks include environmental reconnaissance, armed reconnaissance (locating and attacking targets of opportunity), coastal patrol and interdiction, target and threat assessment, and poststrike reconnaissance.

Information Operations (IO)

IO refers to actions taken to affect adversary information and information systems while defending one's own information and information systems.

Civil Affairs (CA)

CA facilitates military operations and consolidates operational activities by assisting commanders in establishing, maintaining, influencing, or exploiting relations between military forces and civil authorities, both governmental and nongovernmental, and the civilian population in a friendly, neutral, or hostile area of operation. This mission, and the following two missions, are those of SOCOM and not necessarily MARSOC.

Counterproliferation of Weapons of Mass Destruction (CP)

CP refers to the actions taken to seize, destroy, render safe, capture, or recover weapons of mass destruction (WMD). SOF provides unique capabilities to monitor and support compliance with arms control treaties. If directed, SOF can conduct or support SR and DA missions to locate and interdict sea, land, and air shipments of dangerous materials or weapons. SOF is tasked with organizing, training, equipping, and otherwise preparing to conduct operations in support of U.S. government counterproliferation objectives.

Psychological Operations (PSYOP)

PSYOP induces or reinforces foreign attitudes and behaviors favorable to the originator's objectives by conducting planned operations. These operations convey selected information to foreign audiences in order to influence their emotions, motives, objective reasoning, and ultimately the behavior of foreign governments, organizations, groups, and individuals.

IRREGULAR WARFARE (IW)

Irregular warfare (IW) is another mission area for SOF. Unconventional warfare, counterterrorism (CT), counterinsurgency (COIN), civil-military operations (CMO), civil affairs (CA), psychological operations (PSYOP), and foreign internal defense (FID) are all traditional IW activities and core tasks for SOF. With IW's emergence as a focus area for broader participation across the spectrum, it increasingly describes activities that both SOF and general purpose forces will employ in their operational approaches. IW doctrine calls for a suite of capabilities to prevail against those who threaten the United States. IW is a logical, long-term framework that assists in both analyzing and applying many elements of national and international power to achieve mutual security objectives. IW often employs indirect operations to gain asymmetric advantage over adversaries.

CHAPTER 2
HISTORY OF MARINE CORPS SPECIAL OPERATIONS

MARINES ARE NO STRANGERS to the unconventional element of warfighting. The very nature of the Corps, the training and missions, elevates the Marines to the ranks of an elite. On occasion throughout the course of history, these select few have risen beyond the conventional warfighting role and prepared for and executed operations on par with any special operations unit past or present.

Less than twenty years after its war for independence, the fledgling United States of America would wage its first conflict on foreign soil at the beginning of the nineteenth century: the Barbary War (1801–1805). The North African Barbary coast was comprised of four states, nominally ruled by the Turkish Ottoman Empire but in practice given considerable autonomy in which to conduct their affairs: Morocco, Algiers, Tunis, and Tripoli. Sailing in the region were the Barbary Pirates, also known as the Barbary Corsairs, who roamed the waters taxing vessels, attacking ships, and capturing their crews. These Muslim pirates operated much like the Somali pirates of today, but with the sponsorship and support of their Arab rulers.

In the years following American independence, the major European powers paid tribute to the Barbary States in return for their ships being left to ply the Mediterranean unmolested, and the United States did the same. By 1801, however, the United States was behind in its payments, and in that year the bey (ruler) of Tunis, Yusuf Caramanli, declared war in the hope of collecting his tribute by force from the American ships his pirates attacked.

In response, President Thomas Jefferson deployed an armada of frigates, including a Marine expeditionary force, into the Mediterranean in order to defend America's interests in the region. By 1803 the Navy ships had established a blockade of all ports on the Barbary Coast, and its warships were attacking the Barbary Pirate vessels on a regular basis. As in all wars, there are times when the tides shift and the enemy goes on the offensive. Such was the case in October 1803, when the pirates captured the USS *Philadelphia*, which had gone aground while patrolling in Tripoli harbor. The pirates captured the ship, and her captain, William Bainbridge, along with all the officers and crew, was taken ashore and held hostage. The pirates then turned the guns of the frigate against the other U.S. ships. In February 1804, Navy lieutenant Stephen Decatur led a small contingent of eight U.S. Marines in a covert mission to attack the captured ship. Overpowering the guards the men set fire to the ship, denying its use as

OPPOSITE: Marine Raiders sport the two-piece "duck hunter" or "leopard spot" camouflage. The material was printed with a green pattern on one side and a brown one on the reverse side for jungle and beach operations, respectively. The Marines soon found the plain olive color to offer better concealment than either pattern. The latter uniforms were made of cotton herringbone twill in sage green. On the left breast pocket was the Eagle, Globe, and Anchor emblem with "USMC" stenciled above the symbol. NATIONAL ARCHIVES

First Lieutenant Presley O'Bannon led the first Marine attack on foreign soil during the Barbary Coast War. Travelling through the Libyan desert for six hundred miles allowed the Marines and their mercenary allies to take the city of Derna by surprise. Upon taking the harbor fort, O'Bannon ran the colors up the flagpole, and for the first time in history, the stars and stripes of the U.S. flag flew victoriously over foreign soil. USMC ARTWORK

a gun battery to the enemy. British vice admiral Horatio Nelson reportedly stated, "[the raid] was the most bold and daring act of the age."

During this same time, further acts of heroism occurred that would secure the Marines a place not only in military history, but also in the realm of unconventional warfare. On March 8, 1805, 1st Lt. Presley O'Bannon led the Marine detachment consisting of a sergeant and six privates in an unprecedented operation to attack the fortified Tripoli city of Derna and rescue *Philadelphia*'s crew. The Marines linked up with a diplomat turned Navy lieutenant, William Eaton, along with five hundred Greek, Arab, and Berber mercenaries. The coalition force began its mission in Alexandria, Egypt, and traversed six hundred miles of the Libyan desert to assure it the element of surprise.

On March 27 the force reached the city of Derna, where they proceeded to carry out a two-pronged attack; Eaton's force attacked the governor's castle, while O'Bannon led his Marines in a frontal assault on the harbor fort. Braving a hail of musket balls from the defenders, O'Bannon and his men managed to overrun the enemy positions, forcing them to leave with the fort's cannons primed and unfired. The Marines took control of one of the batteries and turned the guns upon the enemy. It was then when O'Bannon placed the American flag upon the ramparts of the fort. After two hours of hand-to-hand fighting, the stronghold was occupied, and for the first time in history, the flag of the United States flew over a fortress of the Old World. The Tripolitans attempted to counterattack the fort several times, yet each time they were repelled, sustaining heavy losses in the process.

The Marines' victory helped Hamet Caramanli, Yusuf's deposed brother, reclaim his rightful throne as ruler of Tripoli. In gratitude, he presented his Mameluke sword to Lieutenant O'Bannon. This famous sword became part of the officer uniform in 1825, and remains the oldest ceremonial weapon in use by U.S. forces today. Derna was the Marines' first battle on foreign soil. Lieutenant O'Bannon and his men are immortalized in the "Marines' Hymn": *"From the Halls of Montezuma to the shores of Tripoli, we fight our country's battles in the air, on land and sea."*

By the late 1890s, Spain had lost much of its power in the world, with the exception of Cuba and Puerto Rico in the Caribbean, and the Philippines and Guam in the Pacific. The sinking of the USS *Maine* in Havana Harbor, Cuba, on February 15, 1898, triggered the start of the Spanish-American War. The Marines, through their actions at Guantanamo Bay and Guam, proved that a small

Marine Raiders run through an obstacle course as they prepare to go to war. They are armed with an assortment of weapons, including the M1 Garand rifle, Thompson submachine gun, and Browning automatic rifle (BAR). USMC PHOTO

Marine Raiders practice beach assaults in a rubber raft. Note the Marine on the bow of the raft providing security with a Browning automatic rifle (BAR). USMC PHOTO

team of men could be rapidly deployed and successfully conduct amphibious operations. By August 12, hostilities had ceased. Though short in duration, the four-month war had transformed the United States into a major world power.

Throughout the late nineteenth century, though their numbers declined, the Marines did battle and protected American interests from Formosa and Korea to the Philippines and China, making use of their expertise in small-unit tactics against the enemy.

As the world entered the twentieth century, the Marines answered the call anew, projecting a U.S. sea-based power whenever the United States needed a rapid response force. The early years saw the Marines active in Central and South America, in what became known as the "Banana Wars."

Their experiences in Central America and the Caribbean led the Marines to begin systematically analyzing the nature and requirements of operations short of war proper or "small wars." Major Samuel M. Harrington of the Marine Corps Schools delivered a formal report, *The Strategy and Tactics of Small Wars*, in 1922. In addition, Major C. J. Miller wrote a 154-page report on the 2nd Marine Brigade's operations in the Dominican Republic titled *Diplomacy and Spurs in the Dominican Republic* in 1923. These and similar Marine publications were inspirational in the first publication of *Small Wars Operations* in 1935. The 1940 version was renamed the *Small Wars Manual*, and remains relevant today as the foundation of a good deal of current thinking and doctrine.

As the twentieth century progressed, the Marines handled an assortment of missions that, by today's standards, would be considered direct action (DA) operations. It would be during World War II, however, that the Corps would first create units specifically trained and equipped with the precise objective to carry out special operations.

MARINE RAIDERS

With the United States fully committed to World War II following the Japanese surprise attack on Pearl Harbor on December 7, 1941, President Franklin D. Roosevelt became interested in the creation of a unit on par with the British Commandos. After some deliberation by Maj. Gen. Thomas Holcomb, commandant of the Marine Corps, he selected the term "Raiders" and created two battalions. The 1st Raider Battalion was activated on February 16, 1942, followed by the 2nd Raider Battalion on February 19, 1942. Lieutenant Colonel Merritt "Red Mike" Edson would command the 1st Raider Battalion and Lt. Col. Evans Carlson commanded the 2nd, thus earning them the names Edson's Raiders and Carlson's Raiders, respectively. Two months later, on April 12, 1942, the Raiders were deployed from the United States en route to American Samoa.

The Marine Raider battalions were designed on similar parameters to today's special operations forces. Each consisted of lightly armed, intensely trained units established by the Marine Corps during World War II. This elite group of Marines would be tasked with three missions: spearhead larger amphibious landings on beaches thought to be inaccessible, conduct raids requiring surprise and high speed, and operate as guerrilla units for lengthy periods behind enemy lines. Having been given the directive from the president, the Raiders enjoyed wide latitude in the acquisitions of weapons and equipment. Whether it was part of the Corps standard issue or not, the Raiders got priority, no questions asked. This included manpower, as the battalions sought after and received the best of the Marine volunteers available.

Edson retained the traditional eight-man squad, made up of a squad leader, two Browning automatic rifle (BAR) men, four riflemen armed with M-1903 Springfield rifles, and a sniper. In contrast, Carlson departed from the norm, organizing his battalion into ten-man teams that included a squad leader and three fire teams of three Marines apiece. These Marines were armed with Thompson submachine guns, BARs, and M-1 semiautomatic rifles.

On August 7, 1942, in the first American offensive of World War II, the 1st Raider Battalion went ashore at Tulagi in the Solomon Islands. It would take three days of intense fighting before the Marines killed all of the Japanese on the island, where the enemy fought to the death. After three weeks, Edson's Raiders were relocated to Guadalcanal

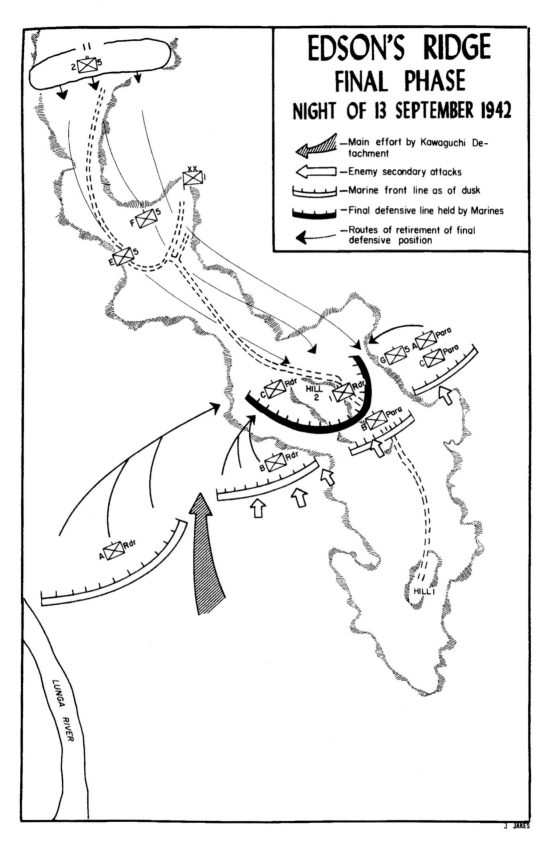

to assist the Marines there in the defense of Henderson Field. At this time, the Raiders continued seek-and-destroy missions against the enemy in a bold raid against approximately one thousand Japanese troops located in the village of Tasimboko. The surprise attack caught the Japanese soldiers off guard, forcing them to abandon their weapons, communications, supplies, and a unit of artillery, all of which the Marines destroyed.

Colonel Edson believed the Japanese forces that had fled during his attack at Tasimboko would attempt to strike Henderson Field. He received the 1st Marine Division's permission to occupy Lunga Ridge located south of Henderson Field in September 1942. The Raider Battalion was reinforced by two companies of the 1st Marine Parachute Battalion.

Edson's Ridge

What was supposed to be a short rest for the Raiders would turn into a fight for their lives. On the evening of September 12, the Japanese unexpectedly attacked their position in an onslaught that included a naval bombardment by a cruiser and three destroyers. This heavy fire forced Edson's force to withdraw to a reserve location. At first light on September 13, Allied aircraft and Marine artillery rained down on the Japanese, forcing them to withdraw from the ridge. Expecting the enemy to attack again, Colonel Edson repositioned his men and instructed them to improve their positions on the ridge.

By this time the Marines were low on ammunition and supplies, but their spirits were high, having been motivated by Colonel Edson, who reportedly told them, "You men

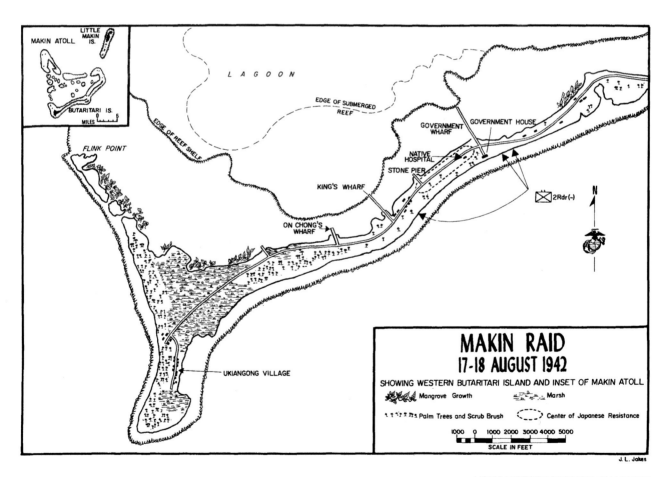

USMC HISTORY DIVISION, QUANTICO

have done a great job, and I have just one more thing to ask of you. Hold out just one more night." And hold out they did, digging in and mustering up the courage, honor, and commitment to battle through the night. More than 2,500 Japanese soldiers repeatedly attacked Edson's Raiders and the Paramarines, whose numbers totaled only eight hundred. When the battle was done, the ground was strewn with the bodies of Japanese soldiers. Colonel Edson and the Marines continued to hold the high ground. The battle that ensued became known as "Bloody Ridge" and eventually as "Edson's Ridge," and has become one of the legendary accounts of courage and tenacity in which the U.S. Marines prevailed despite overwhelming odds during the Pacific Campaign.

Makin Island Raid

On August 17, 1942, Lieutenant Colonel Carlson led the 2nd Raider Battalion on a raid against Japanese forces on Makin in the Central Pacific. The island was triangular in shape and covered in dense coconut palms from the waterline inland. The objective of the mission was to destroy the enemy installation on the island, capture prisoners, and gather intelligence on the Gilbert Island region. The raid was also intended to divert Japanese attention and reinforcements from the Allied landing on Guadalcanal and Tulagi.

The Raiders consisted of two rifle companies and a contingent of the battalion command group. Battalion headquarters, Company A, and eighteen men from Company B (totaling 121 Marines) were embarked aboard the submarine *Argonaut* (APS-1), and the remainder of B Company (totaling ninety Marines) was embarked aboard the submarine *Nautilus* (SS-168). The raiding force was designated Task Group 7.15.

Upon reaching their insertion point at 3:00 a.m., the submarines surfaced, and the Raiders were transferred to rubber rafts. Heavy swells created problems for the Marines as they transitioned from the subs to the rafts, as well as swamping the outboard motors. For these reasons, the original plan of landing the Marines on two separate beachheads was abandoned, and Colonel Carlson ordered the Raiders to follow him to the landing zone, code-named "Z." Fifteen of the eighteen rafts hit the beach along with Carlson, while two others landed approximately a mile north. The last of the rafts landed over a mile to the south, placing the Marines well behind Japanese lines.

Other than the environment obstacle, the Raiders landed unopposed by the occupying Japanese force. Colonel Carlson ordered Company A to head across the island and capture the lagoon road. By 6:00 a.m., the Marines had reached the Government Wharf and reported seizing the objective without enemy opposition. The smoothness of the raid suddenly ended, however, as 1st Platoon, Company A, made contact with enemy forces and met heavy resistance. The Japanese infantry was armed with four machine guns, a flamethrower, grenade launchers, and automatic weapons, as well as snipers. The battle raged on for more than five hours, until another platoon from Company A broke through and flanked the enemy positions.

While Marines had been fighting, a Japanese transport ship and patrol boat had entered into the lagoon. These two ships would never pose a threat to the Raiders, however, as both enemy vessels were sunk by the deck guns of the American submarines on station. Two hours later a flight of twelve Japanese aircraft arrived, consisting of two Kawanishi H6K flying boats, four Nakajima A6M2-N fighters (Zeroes), four Kawanishi E7K2 Type 94 reconnaissance bombers, and two Nakajima E8N2 Type 95 recon planes. The flying boats landed to deliver reinforcing troops, only to be destroyed by gunfire from the Raiders. The other enemy planes flew over the island for more than an hour, sporadically dropping bombs and making strafing runs but failing to kill any Marines.

At 7:30 p.m., as scheduled, the Raiders began their withdrawal from the island, using the same eighteen rafts on which they had come. Though the Marines had practiced surf passages and raft operations extensively, the heavy sea conditions washed many of the rafts back to shore. Only seven of the rafts made it back to the submarines; the balance of the Raiders remained on the island. On the morning of August 18, the submariners dispatched a rescue raft to deploy a rope that would allow the Marines to be pulled out through the rough surf. However, a group of Japanese aircraft arrived and began attacking the waterborne Americans, forcing the submarines to crash-dive and remain submerged for the balance of the day.

The Raiders were able to get a signal to the submarine and arrange a new rendezvous for 11:00 p.m. at the entrance of Makin Lagoon. The Marines did what Marines still do today: they adapt, improvise, and overcome. Using their four remaining rafts and two native outriggers, the Raiders headed out to sea at 8:30 p.m., arriving alongside the *Nautilus* at 11:08 p.m., just off the entrance to the lagoon, Flink Point. Seventeen of the Raiders were wounded and thirty-one others remained unaccounted for. It was later learned that nine of them were still alive on the island after the last evacuation, were captured by the Japanese and transported to Kwajalein in the Marshall Islands. There, Rear Admiral Koso Abe, commander of the 6th Base Force, ordered them beheaded.

Although the Raiders accomplished their mission of decimating the Japanese garrison on Makin Island—Japanese dead were estimated at anywhere between 83 and 160—the overall evaluation was that it had failed to meet the strategic goals of the operation. No enemy prisoners had been taken, and no workable intelligence collected. The raid did prove useful, however, in testing the Raiders' tactics, techniques, and procedures.

Two additional Raider battalions would be created and serve operationally in the Pacific theater of operations. The Raiders would participate in twenty major campaigns and battles in the course of the war, bringing their fighting spirit and discipline to every engagement. Some Marines viewed the Raiders as an "elite within an elite," which lead to some resentment in the Corps. As such, the majority of operations conducted by the Marine Raiders were seen as tactics that should be employed to all Marine infantry, and all of the Corps should be trained to carry out the same tactics. By February 1944, the Marine Raider battalions had been disbanded.

Paramarines

In May 1940 the commandant of the Marine Corps, General Holcomb, ordered plans for the creation of a Marine unit capable of deploying via parachute. This unit would consist of one battalion of infantry and a platoon of artillery consisting of two 75mm pack howitzers. The plan outlined three tactical mission parameters under which the Marines would be deployed: First, as a reconnaissance

A Paramarine armed with an M1911 .45-caliber pistol and Reising submachine gun. An air-cooled, delayed blowback, shoulder-fired weapon, the Reising weighed 6.5 pounds and had a maximum effective range of 300 yards. It was chambered for .45-caliber ACP ammunition in either a twelve- or twenty-round magazine, and had a muzzle velocity of 900 feet per second with a cyclic rate of 450 rounds per minute. Unique to the Marine Corps, the Reising was deemed problematic, and by the spring of 1943, the Marines replaced it with the Thompson SMG or M1 carbine. COURTESY OF JOE PAGAC

and raiding force that would operate under the assumption that it would potentially be unable to return to Allied lines. Second, the paratroopers could be deployed as a spearhead or vanguard to capture and hold strategic positions or terrain until relieved by larger follow-on forces. Third, the unit could act as an independent force functioning in a guerrilla warfare role behind enemy lines.

On October 26, 1940, the first group of two officers and thirty-eight enlisted men were sent to the Naval Air Station in Lakehurst, New Jersey. Here these forty Marines would begin their training, learning how to jump out of an airplane. Whether the motivation was patriotism or the lure of the extra jump pay—fifty dollars per month for enlisted and a hundred dollars per month for officers—there was no shortage of volunteers. Marines trying for a slot in the parachute skill were required to be between eighteen and twenty-two years old, single, athletic, and intelligent, with no physical or mental impairments.

The tempo of Marine training was beginning to feel the demands of the war effort. For this reason the secretary of the Navy, William Franklin Knox, approved the Marine commandant's request to establish two Marine parachute schools. In May 1942 the Parachute School was created in San Diego, and in June the Parachute School opened at New River, North Carolina.

The 1st Parachute Battalion officially formed at Quantico on August 15, 1941, under the command of Capt. Marcellus Howard, and on July 23, 1941, the 2nd Parachute Battalion was organized at San Diego under the command of Capt. Charles Shepard Jr. The 3rd Battalion was formed on September 16, 1942, under the command of Maj. Robert Vance, and the 4th Parachute Battalion entered on April 2, 1943.

The 1st Parachute Battalion was attached to the 1st Raider Battalion, along with the 2nd Battalion/5th Marines, which made up the Marine forces landing on Guadalcanal. The 1st Parachute Battalion, which had suffered heavy casualties in the Battle of Tulagi and nearby Gavutu-Tanambogo in August, was placed under Colonel Edson's command. The 2nd Parachute Battalion would be tasked with a diversionary raid on Choiseul Island in October 1943, and later joined the 1st and 3rd Parachute Battalions on Bougainville.

In retrospect the tactical advantage of airborne troops cannot be discounted in either a full-scale war or low-intensity conflict. Nevertheless, the Paramarines were considered a luxury item the Marine Corps could neither fund or support logistically at that time in history. On February 29, 1944, the 1st Marine Parachute Regiment was disbanded.

Office of Strategic Services (OSS)

When one mentions the OSS, the average person may think of the historical link to the U.S. Army and Special Forces, which evolved from the Jedburgh teams and operations groups. The OSS, which was the forerunner of today's Central Intelligence Agency (CIA), ran clandestine missions in both the European and Pacific theaters of operations. It was instrumental in gathering intelligence, conducting sabotage raids in occupied Europe, and working with French resistance fighting against the Nazis.

What may be surprising to some is that Marines were also active in this covert unit during the war. Assigned to the OSS was Marine lieutenant colonel William Eddy, who worked with the counterintelligence (CI) branch within the Office of Naval Intelligence prior to the outbreak of World War II. The head of the OSS, Army colonel William "Wild Bill" Donovan, assigned Eddy to head all undercover and subversive operations in French North Africa. Eddy coordinated efforts between the OSS, Special Operations Executive (British sabotage and subversion agency), and the British Secret Intelligence Service (SIS) for the Allied landings called Operation Torch. Starting on November 8, 1942, Torch would be the first military action in which the OSS provided direct support.

Another Leatherneck to operate with the OSS was Col. Peter Ortiz, then a captain, who served in both Europe and Africa. Ortiz was born in New York City on July 5, 1913, but was schooled in France, where he left prior to his graduation in order to join the French Foreign Legion. He was wounded in action and captured by the Germans, but after fifteen months as a prisoner of war (POW), he managed to escape in October 1941. Making his way back to America, he joined the Marines on June 22, 1942, and was trained at Paris Island. In May 1943 he was attached to the OSS.

Having lived in France, Ortiz was fluent in that language as well as five others; he was also a graduate of the Marine

Sixteen members of a reconnaissance company attached to the 1st Marine Division in Korea reveal their position. USMC HISTORY DIVISION, QUANTICO

A Scout patrol of the Marine Amphibious Reconnaissance Company departs in rubber boats from the submarine USS *Perch* during Operation Ski Jump, January 1957. USMC HISTORY DIVISION, QUANTICO

parachute school. On January 6, 1944, he parachuted into France to assist the underground resistance forces known as the Maquis. While working with the Maquis, he also assisted in the rescue of four Royal Air Force pilots who had been shot down. In 1944, Ortiz was recaptured by the Germans and spent the remainder of the war as a POW.

Amphibious Reconnaissance Company

On January 7, 1943, the Observer Group, which had been created in 1942, was redesignated the Amphibious Reconnaissance Company under the command of Capt. James L. Jones. In the summer of 1943, the Amphibious Corps would be reorganized, and the ARC would become V Amphibious Corps, or VAC Amphib Recon Company.

During operations on Apamama, Majuro, and Eniwetok, VAC Amphib Recon Company was assigned to pinpoint the location of the enemy throughout the atolls. The VAC Amphib Recon Battalion underwent further evolution on April 14, 1944, when it became the Amphibious Reconnaissance Battalion with a complement of twenty officers, two hundred seventy enlisted Marines, and thirteen Navy corpsmen. Worth noting is that at the same time, Marines were attached to the "coast watcher" units, and recon teams from "Special Service Unit Number 1" served the forces in the Southwest Pacific.

A pivotal moment for the Reconnaissance Marines would come in July 1944 when Marine Lt. Gen. Holland Smith and Vice Adm. Kelly Turner disagreed on the

Members of the 1st Marine Division's Reconnaissance Company review the results of a night patrol in Korea. Examining a map are, left to right, Capt. James H. A. Flood, Sgt. Richard P. Antekeier, and Hospital Corpsman 1st Class Ernest A. Colburn. USMC HISTORY DIVISION, QUANTICO

A parachute scout, foreground, makes a sketch of enemy terrain and installations while another Marine Corps scout covers him with an M3 "grease gun." All reconnaissance Leathernecks are experts in determining terrain factors and the capabilities of roads and bridges. These Marines are from the 1st Marine Division, Camp Pendleton, California. USMC HISTORY DIVISION, QUANTICO

proposed landing sites for the assault on Tinian Island. Armed with only Ka-Bar knives, the Marines of the Amphibious Reconnaissance Company, VAC, were tasked with surveying two beaches that Smith had chosen as possible locations for the landing.

Brigadier General Russell Corey (Ret.) relates: "There were two landings sites, White 2 and White 1. The beaches were about sixty yards and one hundred sixty yards respectively. [Corey] and Gunny Charles Patrick (Major, Ret.) led the two recon teams. The first night the current was so strong we could only conduct the recon on one of the beaches. The next night we returned and were able to do both landing sites."

Based on VAC's findings, the site choice went to General Smith, and beaches White 1 and White 2 were chosen for the landing of two Marine divisions on Tinian. General Corey adds: "Had the division gone ashore on the spot picked by Kelley, it would have been a slaughter. The Japs had gun emplacements and artillery all lined up in the Tinian Town location." The Recon Marines had earned their pay and would see subsequent use in Iwo Jima and Okinawa, as well as other amphibious assaults in the Pacific.

BIRTH OF FORCE RECON

Force Reconnaissance, as it is known today, was activated on June 19, 1957, with the creation of 1st Amphibious Reconnaissance Company, FMFPAC (Fleet Marine Force, Pacific), under the command of Maj. Bruce F. Meyers. Located out of Camp Pendleton, California, this newly organized company would be formed into three platoons: an amphibious reconnaissance platoon,

Clad in a rubber exposure suit, a Marine scout swimmer dives from an inflatable boat more than five hundred yards offshore. The exposure suit is worn in cold climates where water temperatures hamper swimmers. Working in teams of two, the scout swimmers examine the beach for obstacles and enemy patrols. If the area is clear, they signal the remainder of the reconnaissance patrol to come ashore. USMC HISTORY DIVISION, QUANTICO

A patrol from B Company, 3rd Recon, 3rd Marine Division, searches an area just below the Vietnamese Demilitarized Zone (DMZ), January 1969. USMC HISTORY DIVISION, QUANTICO

Cpl. John L. Borst and Staff Sgt. David D. Woodward, both with the 3rd Force Reconnaissance, call on the radio for a gunship to hit the area ahead of them while on patrol "touchdown" near the DMZ and the Laotian border, February 1967. USMC HISTORY DIVISION, QUANTICO

a parachute reconnaissance platoon, and a pathfinder reconnaissance platoon. Subsequently, in 1958, half of the company was transferred from Camp Pendleton to Camp Lejeune, to form the 2nd Force Reconnaissance Company, FMFLANT, (Fleet Marine Force, Atlantic), under the command of Capt. Joe Taylor, and thus supported the 2nd Marine Division. Worth noting is the fact that it would be another four years before the Navy SEALs would come on the scene, and another eleven years before the Army would designate a counterpart to Force Recon with the creation of LRRPs (long-range recon patrols).

As the country progressed into a new decade, the Marines of Force Recon also advanced in their tactics, techniques, and procedures. These men, proficient in land navigations, small arms tactics, and patrolling, as well as having earned their wings, would now depart from terra firma to master the skills of attacking from the sea. Fast boats, rubber rafts, and submarines were not new to the Marines, who had used these craft as platforms during World War II and Korea. Methods of locking-in and –out of submarines and buoyant ascents all would serve the Force Recon Marines in performing their missions. While the men were honing their skills, the United States was becoming increasingly involved in the guerrilla war in the Republic of Vietnam. That conflict would prove to be a watershed for Marine Force Reconnaissance.

Sergeant Ron Hillard from Birmingham, Alabama, looks out the porthole of a CH-46 helicopter as he waits for his team to be inserted, May 1970. USMC HISTORY DIVISION, QUANTICO

The Vietnam War

Members of 1st Force Reconnaissance Company were deployed to Vietnam in 1965, while 2nd Force had the assignment of training new Recon Marines to be sent to Southeast Asia. The Force Recon Marines at Lejeune would also serve as the primary unit should any other contingency arise elsewhere that required the attention of the Marine Corps. Sensing the need for additional reconnaissance capabilities in country, the Corps formed 3rd Force Company, per Lt. Col. Bill Floyd (Ret.): "We took the company over in September of 1965. First would make their home in Dong Ha, while 3rd was in Quang Tri." When the Marines of Force Recon first deployed to Vietnam they were armed only with M3A1 "grease guns"; subsequently they would reequip with the M-14 and eventually with the M16 assault rifle.

Marines are taught to fight, to give no quarter, and to defeat the enemy at all costs. Conversely, reconnaissance work required stealth and patience, at times letting the enemy pass by so the team could report on their movements. The aggressiveness of the U.S. Marine was something that had to be "un-learned" in order for reconnaissance to be successful. To enhance these new skills of Force Reconnaissance companies, many of the Marines attended Recondo School taught by members of the U.S. Army 5th Special Forces Group (Airborne).

Force Recon and Battalion Recon Post Vietnam

Worth noting is that within the Marine Air-Ground Task Force there are two reconnaissance units: one of these is Force Reconnaissance Company, while the other is the Battalion Recon (sometimes referred to as Division Recon) platoon, which is under the commander of the combat element. The Force Recon Company works directly for the MEF commanding general, normally a three-star, or directly for the MEU (SOC) commander, providing tactical and strategic reconnaissance and limited scale raids (e.g., GOPLAT, VBSS, demolition raids, etc.) in the deep battlespace. Force Recon covers the commander's area of *interest*: the edge of the artillery fan and beyond. The Battalion Recon covers the commander's area of *influence*: within the artillery fan (approximately thirty kilometers from the forward line of troops).

Armed with an M4A1 carbine, a Marine provides security for his team as the platoon leader calls in a SitRep. The 4th Force Reconnaissance Company is located at the Marine Corps Base, Kaneohe Bay, Hawaii.

The Force Reconnaissance Company will conduct operational-level reconnaissance in the deep battle areas. Conversely, Battalion Recon Marines are trained to operate just forward of the front lines, or directly in front of or alongside of the conventional Marine units. Not all Battalion Recon Marines are airborne and/or scuba qualified. While the training paths are similar for the Recon Marines, those assigned to Force Reconnaissance companies will receive more advanced skills and training, which more than match those of the special operations forces of SOCOM. This skill set allows the Force Recon Marines to excel in the area of deep reconnaissance and DA missions, well beyond the range of the Battalion Recon assets. The special skills of these Marines, from Lt. Presley O'Bannon to the Marines of today, create a tight link between those historical units and the Marines of MARSOC.

CHAPTER 3

MARSOC CREATION AND ORGANIZATION

IN 1992, THE SOCOM/USMC board was established in an attempt to formalize the joint efforts of SOCOM and the Marine Corps; it was terminated four years later. After al Qaeda's murderous airliner hijackings and suicide attacks on the World Trade Center and the Pentagon on September 11, 2001, the U.S. military machine went into overdrive. Operational Tempo, or OPTEMP, would go through the roof, especially among special operations forces. While it is the soldier who prays most for peace, this was the war for which SOCOM units had been trained, equipped, and prepared to fight. SOCOM units from Tier-1 (i.e., Delta Force) to support elements were "jocked up" and poised for *their* rendezvous with destiny. This was the Super Bowl of missions for the special operations community, and each operator was primed and ready to answer the call of the commander in chief, to "bring the enemy to justice, or justice to the enemy."

The global war on terrorism (GWOT) resulted in the largest deployment of special operations forces since the Vietnam War. Although the Marine Corps did not have any troops under SOCOM, it did have the MEU (SOC), and organic to the force were the Force Recon Marines. These Marines received the same training as the SOCOM units and, without doubt, were on par with most, and head and shoulders above some of those in the SOF community. Force Recon Marines were so highly regarded for their special operations skills that it is said that Delta Force assets aboard the USS *Kitty Hawk* during Operation Enduring Freedom wanted Marine Force Recon teams to conduct their preassault reconnaissance.

It soon became apparent that SOCOM had to increase both in capability and capacity in order to meet and prosecute the demands of the GWOT. In November 2001, Commandant Gen. James L. Jones and the commander of SOCOM, Air Force general Charles R. Holland, revived the SOCOM/Marine board. The intent was to move the Marine Corps and SOCOM closer together and establish a framework for building bridges between the two organizations. A year later, in December 2002, Marine Corps Bulletin 5400 called for a "proof-of-concept" operation for a Marine Corps contribution to SOCOM. The test period would last for two years, beginning in 2004; preparation and training, however, had already commenced in 2003.

A memorandum of agreement (MOA) between SOCOM and the Marine Corps was signed off in February 2003 by Lt. Gen. Bryan D. Brown, U.S. Army, deputy commander, SOCOM, and Lt. Gen. Emil R. "Buck" Bedard, deputy commandant for plans, policies, and operations (PP&O), USMC. The agreement set up the parameters for the proof-of-concept period for the Marine Corps' participation in SOCOM.

OPPOSITE: U.S. Marines provide security in a wheat field in Suji as Afghan soldiers and Marines patrol through the village in the Bala Baluk district in Afghanistan's Farah province on March 29, 2010. The Marines are assigned to Marine Special Operations Command. USAF PHOTO, STAFF SGT. NICHOLAS PILCH

ABOVE: Detachment One headquarters at Camp Del Mar Boat Basin, Camp Pendleton, California. Marine Corps Special Operations Command Detachment One (MCSOCOM Detachment One or Det 1) was a proof-of-concept program to assess the feasibility of Marine Special Operations Forces being permanently detached to the United States Special Operations Command (SOCOM). It was commanded by Col. Robert J. Coates, former commanding officer of 1st Force Reconnaissance Company.

LEFT: Created by Warrant Officer (then Gunnery Sgt.) Anthony Siciliano, Det One's insignia comes from the World War II Marine Raider's patch, a blue patch with a skull and stars. The scarlet, blue-and-gold disk represents the unit's joint Navy-Marine Corps origins. The crossed stiletto/lightning bolt represents the unit's special operations mission and its global communications reach. The parachute wings represent airborne qualified status, and the mask above it represents the combat diver qualification. USMC ARTWORK

A pair of Marines from the Intel section of Det One prepare for a patrol. Both Marines are armed with the Colt M4A1 carbine, with an assortment of COTS and issued add-ons. The Marine on the left has kept his carbine black, while his teammate has painted his weapon in a desert camouflage pattern. Both are carrying M1911 pistols in a Safariland Model 6004 tactical holster. Adding to the "cool" factor are the sunglasses, a.k.a. "tactical eye protection." MARSOC/DET ONE PHOTO

The official name of the test unit was USMC/USSOCOM Detachment (MCSOCOM Det). The detachment would work with a deployed naval special warfare (NSW) squadron to augment the squadron's capabilities in special reconnaissance, direct action, and other missions. It would be under the operational control of an NSW commander while attached to the NSW squadron. In addition to a headquarters section, MCSOCOM Det would have sections specializing in reconnaissance, intelligence, and fire support. The section personnel would work together in teams organized to fit the demands of specific missions. The MOA emphasized that detachment integrity should be maintained to the greatest possible degree, although it allowed for the NSW squadron commander to use small teams of Marine detachment personnel with his other squadron assets if the mission required it. And—very important when it comes to the battles of the budgets—the Marine Corps would bear all personnel, equipment, facilities, training, deployment, and operational expenses required to support the detachment.

Out of the above MOA came the T/O (table of organization) for Marine Corps Special Operations Command Detachment One (MCSOCOM Detachment One or Det 1), officially activated at Camp Pendleton, California, on June 20, 2003. The unit was commanded by Col. Robert J. Coates, former commanding officer of both the 1st Force Recon Company and Marine Special Operations Training Group (SOTG), located at Camp Pendleton. Det One was headquartered at Camp Del Mar Boat Basin, located in Camp Pendleton.

A member of Det One waits to move out on a night patrol in Iraq in support of Operation Iraqi Freedom. Det One Marines had wide latitude in modifying their weapons with COTS equipment. This Marine has replaced the standard Colt furniture on his M4A1 with a Magpul M93 modular stock system, as well as an Ergo brand pistol grip. Atop the carbine is an Aimpoint 'red dot' sight, AN/PEQ-2, and Surefire flashlight. MARSOC/DET ONE PHOTO

ORGANIZATIONAL STRUCTURE OF MCSOCOM DETACHMENT ONE

Each MCSOCOM detachment consisted of eighty-one Marines and five Navy corpsmen divided among four sections: a reconnaissance element (thirty men); intelligence element (twenty-nine men) containing a headquarters element and a radio recon team (RRT) (nine men), a human intelligence (HUMINT) exploitation team (HET) (six men), and all-source fusion team (twelve men), which assembled information from various sources, then analyzed and distributed it to all who needed it; a fire support unit (seven men) composed of three ANGLICO (air-naval gunfire liaison company) Marines, three radio operators, and a forward air controller, led by a field artillery major; and a headquarters element.

Det One was designed to operate as an independent organization commanded by a colonel (O-6) with a major (O-4) as the deputy. The headquarters element consisted of staff sections representing operations, logistics, communications, etc. Once deployed, because of the previously agreed upon command arrangements, the Det

One commander was detached from the detachment and attached to 1 MEF to conduct liaison and coordination with conventional Marine Corps operations, while the deputy assumed command of Det One. Designated officers and NCOs from the detachment were then integrated with the leadership of the Naval Special Warfare Task Group that they were co-deployed with. The recon element was led by a captain (O-4) who had been an infantry platoon commander with reconnaissance background. The platoon sergeant of the recon element was a master sergeant (E-8) with extensive reconnaissance background. The recon element included four five-man teams.

Colonel Robert J. Coates, the detachment's founding commander, related, "The task for the day was training, training and training." And train they did. Within eight months the detachment went from scratch to a combat unit ready to deploy to war. Colonel Coates's requirements were straightforward. "The operator needed hard feet, a strong back and [to] be a gun fighter." The average age of a Marine in Det One was thirty-six to forty-one.

From April through October 2004, Det One would be deployed to Iraq in support of Operation Iraqi Freedom. The Marines operated as Special Operations Task Unit Raider under the control of the Combined Joint Special Operations Task Force Arabian Peninsula (CJSOTF-AP). The detachment operated with Naval Special Warfare Group One to execute direct action, coalition support, and battlefield-shaping operations.

The proof of concept for Det One posed several questions and criteria for the Marines' future within SOCOM as the queries included: Can a Det One operate effectively as a stand-alone special operations force (SOF) component under a Joint Special Operations Task Force (JSOTF)? What SOF core tasks can Det One accomplish? Can Det One perform effectively in the full spectrum of operational environments?

A proof-of-concept deployment evaluation of Det One's actions during Operation Iraqi Freedom prepared by the Joint Special Operations University at Hurlburt Field, Florida, states: *"Employment in OIF was limited to DA,*

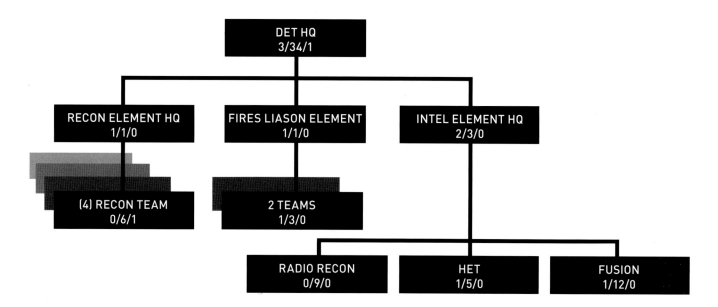

Marine Corps Detachment Task Organization in Iraq
TU Raider ISO NSWTG-AP

SR and support to Special Activities. This limited scope of employment was based on requirements and priorities of CJSOTF-AP and NSWTG-AP commanders in OIF. Thus it was an externally imposed constraint that precluded the MCSOCOM Det from executing other core tasks. Based on the operational effectiveness demonstrated by the detachment in the assigned tasks, it is reasonable to suggest that given similar preparation, the Det could satisfactorily conduct or support other core tasks.

"The MCSOCOM Det demonstrated they could effectively conduct Direct Action (DA) and Special Reconnaissance (SR) in conjunction with a Naval Special Warfare Task Group. Given the personnel qualifications, training and equipment, of the Detachment it is reasonable to suggest they could also conduct or support Foreign Internal Defense (FID), Counter Terrorism (CT), Special Activities, Support to selected Theater Security Cooperation Plans (TSCP), and other tasks as required.

"The MCSOCOM Det conducted 23 Direct Actions (DA) Raids. The raids can be characterized as well planned and executed and complex in nature. All but one of the raids were conducted with support from one or more fires platforms and at least one other CSOTF-AP ground force. 85% of the targets raided yielded detainees or contraband, and one out of every 12 detainees were HVTs. Of note, 9 insurgent cell leaders were detained or killed (1 was killed), and 7 Improvised Explosive Device cells were disrupted. Research and analysis strongly indicate the initial force contribution was an overall success and that a Marine Corps force contribution has the potential to support USSOCOM requirements."

With the global war on terrorism drawing more and more on special operations forces, the decision was made to officially incorporate a Marine Corps special operations force element into SOCOM. In an announcement by Secretary of Defense Donald Rumsfeld on November 1, 2005, he approved a joint recommendation by SOCOM and the Marine Corps to add Marine Corps special operations forces to SOCOM. Det One had proven its worth; this was the birth of MARSOC. The Marine Corps now joined America's other armed forces as a full-fledged member of the special ops community.

During a press conference, Rumsfeld stated: "In this complex and unconventional conflict, we are constantly looking for ways to strengthen our armed forces. For a number of years, the chairman [of the joint chiefs] and I and others have been working on plans to further improve the special forces. One of the results of those studies is that I've just approved the creation of a Marine Corps component in the U.S. Special Operations Command, which will increase the number of Special Operations forces available for missions worldwide while expanding their capabilities in some key areas."

In October 2005, the Foreign Military Training Unit (FMTU) was activated under the 4th Marine Expeditionary Brigade (Antiterrorism). The mission of the FMTU was to train in combat advisor techniques, language and culture, and mission-specific skill sets, depending on the requirements of the host nation. FMTU Marines are also required to be duty experts in infantry skills, combat medicine, and advanced communications, because they are often working in small teams independent from friendly forces, and in some cases may be the only U.S. military forces in a foreign country. In May 2007, FMTU was changed to the Marine Special Operations Advisory Group (MSOAG) and deployed to Afghanistan in support of Operation Enduring Freedom. On August 27, 2009, MSOAG was redesignated as the Marine Special Operations Regiment (MSOR). On March 10, 2006, Det One was deactivated and succeeded by MARSOC.

MARSOC

On February 24, 2006, the Marine Corps officially joined the special operations community with the creation of MARSOC. MARSOC became a major command within the Corps and SOCOM. At the activation ceremony of this important milestone in the history of the Corps, Secretary of Defense Rumsfeld stated, "It pairs two of history's most dedicated groups of warriors—the men and women of the U.S. Special Operations Command and the United States Marine Corps. Special operations forces and U.S. Marines are legendary for their agility, creativity and willingness to take on difficult missions,

Assuming command of the Marine Special Operations Command on February 24, 2006, Brig. Gen. Dennis J. Hejlik was a former enlisted Marine who had been discharged as a sergeant in 1972 and commissioned through the Platoon Leaders Class Program. He is a graduate of the Amphibious Warfare School, the Command and Staff College, and the Naval War College. USMC PHOTO

and the Marines have played important roles in past U.S. victories. Today in the global war on terror, we call on Marines again . . . to seek new and innovative ways to take the fight to the enemy. Our country needs agile, highly mobile forces to track down terrorist cells that are dispersed across the globe."

The backbone of MARSOC came from the 4th Marine Expeditionary Brigade (Antiterrorism) headquarters. Additionally, the FMTU and several Force Reconnaissance platoons were reassigned to create the new Marine special operations companies (MSOC). In 2010, the Marine commandant, Gen. James Conway, commented, "We had to pay the upfront price of becoming involved in true special operations type of capabilities, 2500 of some of our absolute best Marines, and we want to sustain that over time." As of February 28, 2010, there were 1,965 Marines, 180 sailors, and 124 civilians assigned to MARSOC, a total of 2,269 personnel.

According to Maj. Jeff Landis, MARSOC PAO: *"MARSOC's role within SOCOM has expanded in that MARSOC is better able to assist SOCOM in stabilizing fragile governments so that they can defend their sovereignty, not only territorially, but economically and culturally as well. MARSOC does this by maintaining an enduring 1-SOTF [special operations task force] and 2-MSOC [Marine special operations company] presence while still allowing the command the flexibility to support additional SOCOM requirements as they arise. In our ongoing efforts to build a command with long-term relevancy that USSOCOM can employ across the spectrum of SOF engagement methods, the top priority remains in selecting and training Marines into regional experts who can operate in an area and work by, with, and through the native population. We must continue to develop our Marines and sailors [who] will eventually be capable of conducting missions that can help shape the long-term strategic environment.*

"The decision to create a MARSOC comes after several years of considerable effort on the part of SOCOM, the Marine Corps and the Pentagon to build a Marine contribution to a community that has been heavily used since the war on terrorism began. In our ongoing efforts to build a command with long-term relevancy that USSOCOM

The official emblem of the United States Marine Forces Special Operations Command.

can employ across the spectrum of SOF engagement methods, MARSOC's top priority is to select and train Marines who will eventually be capable of conducting missions that can help shape the long-term strategic environment using capabilities to separate reconcilables from irreconcilables. MARSOC brings the Marine Air-Ground Task Force (MAGTF) mentality to SOF.

"The Marine Corps is inherently expeditionary in nature and regularly organizes its units so that they have the ability to work in austere environments with very little reach back requirement. MARSOC incorporates fused intelligence capabilities, combat support (CS), and combat service support (CSS) directly into our units at the lowest level so that they are specifically task-organized and self-sustaining. We simply took the MAGTF mindset inherent in the Marine Corps, compressed it into a smaller team, and trained and equipped that team exclusively and specifically for special operations missions. The Global War on Terror has dramatically expanded our requirement for SOF. MARSOC has created a new group of mature, experienced, and combat-tested men who add both special operations capacity and capability—a strong positive step in protecting our way of life."

MARINE SPECIAL OPERATIONS REGIMENT

The MSOR, located at Camp Lejeune, North Carolina, is made up of three Marine special operations battalions and a headquarters company. According to the operations officer for the MSOR, the changes were made to create three battalions, each with equal capabilities in direct action, special reconnaissance, and foreign internal defense. The regiment has absorbed the tasks of the Marine Special Operation Advisory Group (MSOAG), formally the FMTU. Marines provide specific military training and advisory support to foreign forces, which in turn organize and enhance those countries' tactical capabilities to perform in Joint Special Operations Task Force as directed by SOCOM. The Marines and sailors of the regiment advise, train, and assist the forces of friendly host nations.

These foreign internal defense (FID) activities constitute one of the core missions of SOCOM. The FID missions enable the host nation's maritime, naval, and paramilitary forces to support their government's internal security, and to counter the threats of subversion and insurgency, both internal and external.

The benefits of a FID mission can be manifold: first, you have developed a new ally; second, should the need arise to deploy SOCOM units into the region, the host nation is already trained in similar combat tactics; and third, if the host nation can provide its own security in combating any threat, it would curtail the need to commit massive amounts of U.S. forces into the region.

A pair of MARSOC Marines makes a waterborne insertion, using Draeger LAR-V rebreathers. The closed-circuit self-contained breathing apparatus does not emit any bubbles to alert the enemy to their presence. They are armed with M4A1 carbines equipped with AN/PEQ-2 IR laser sights and suppressors. The Marine on the left has fitted his carbine with an EOTech holographic sight, while the one on the right is using an ACOG optic. Weapons at the ready, they will keep the mouthpiece of their rebreathers in place, in the event they need to make a hasty return to the water. USMC PHOTO

Two Marines with a Marine Special Operations Company are on the lookout for possible Taliban fighters in a village in Helmand province, southern Afghanistan. The MSOC unit and Afghan National Army soldiers were visiting the village in which they had distributed humanitarian aid. USMC PHOTO, STAFF SGT. LUIS P. VALDESPINO JR.

Marines with 3rd Marine Special Operations Battalion, MARSOC, conduct a direct action hit during training at the Mountain Warfare Training Center located in the Sierra Nevadas, northwest of Bridgeport, California. The training was part of a three-week exercise in which the operators sharpened their skills in an environment containing elevations up to 7,500 feet above sea level. USMC PHOTO, SGT. STEVEN KING

Marine Special Operations Battalion

MSOB Marines train in special reconnaissance, sniper skills, specialized insertion and extraction that includes helicopter operations, open and closed-circuit diving, and the full spectrum of airborne operations from static line to military free-fall operations (i.e., HALO/HAHO; close quarters battle, precision marksmanship, and coordinated reconnaissance and direct action mission profiles). Training in these skills is geared to a unique set of special operations conditions and standards. With the activation of the MSOBs, the Marines deactivated the 1st and 2nd Force Reconnaissance companies. In October 2008, by the direction of the commandant of the Marine Corps (CMC), the D companies within both 1st and 2nd Reconnaissance Battalions were redesignated as "Force Reconnaissance" companies and placed under the operational control of the Marine expeditionary force, particularly I MEF and II MEF, accordingly. These companies will assume the traditional deep reconnaissance and supportive arms mission, and will deploy in support of the current directive required by the Marine Corps.

When it positively, absolutely has to be destroyed overnight, send in a Marine with an M107 special application scoped rifle. Wearing a ghillie suit to blend in with his surroundings, this Marine sniper moves stealthily through the woods to his objective. Once on site he may wait hours, perhaps days, to engage his target. MARSOC PHOTO

1st Marine Special Operations Battalion

The 1st Marine Special Operations Battalion (MSOB) is headquartered at Camp Pendleton, California, on the same real estate originally occupied by Det One. The mission of 1st MSOB is to organize, equip, and train its Marines to be ready to deploy for operations around the world as directed by MARSOC. Each of the battalions consists of four Marine special operations companies (MSOCs). Each company in turn is made up of four Marine special operations teams (MSOTs). The MSOT is where the boots hit the ground; it operates with a fourteen-man team, complete with critical skills operators (CSOs—a.k.a. "door kickers"), communications specialists, corpsmen, intelligence, fire support, and other support personnel. MARSOC's organization at the Marine special operations company (MSOC) level is different from other special operations forces (e.g., the Army's SF Operational Detachment Alpha [ODA] operates a twelve-man team, and a SEAL platoon is seventeen men). The Marines integrate a direct intelligence and enabling package within the MSOC to provide enhanced situational awareness for each team.

A U.S. Marine Corps Forces, Special Operations Command Marine and Combined Joint Special Operations Task Force soldier move to a higher position during a route recon patrol through Farah province, Afghanistan. The rugged terrain makes it difficult to hunt down the elusive enemy . . . but not impossible. USAF PHOTO, STAFF SGT. NICHOLAS PILCH

Marines with a Marine Special Operations Company take aim on Taliban fighters in a Helmand province village in late February 2008. Afghan National Army soldiers and the MSOC Marines were visiting the southern Afghanistan village when they were attacked by Taliban fighters. A MK 19 40mm automatic grenade launcher can be seen in the cupola of the ground mobility vehicle. USMC PHOTO, STAFF SGT. LUIS P. VALDESPINO JR.

2nd Marine Special Operations Battalion

The 2nd MSOB was activated on May 15, 2006, and is headquartered at Camp Lejeune. Each of the MSOBs is commanded by a lieutenant colonel, while the MSOCs are commanded by a major. Each of the MSOCs is capable of deploying a task-organized expeditionary SOF team to conduct core SOCOM missions in support of the geographic combatant commanders (e.g., CENTCOM, etc.). On January 4, 2007, Fox Company deployed for real-world operations with the 26th Marine Expeditionary Unit (MEU) in support of Operation Enduring Freedom.

3rd Marine Special Operations Battalion

The 3rd MSOB is also stationed at Camp Lejeune and deploys at the direction of the MSOR. In addition to the Marine operators, it is planned to have at least one special amphibious recon corpsman (SARC) deploy with every team, with an independent duty corpsman (IDC) a step above that. Currently, the MSOTs have a mixture of SARCs and IDCs. Worth noting is these corpsmen, regardless of Navy rating, are active members of the MSOT and are all shooters.

MARINE SPECIAL OPERATIONS SUPPORT GROUP

The MSOSG was activated on January 19, 2007, at Camp Lejeune. The MSOSG equips, trains, structures, and provides specially qualified Marine forces. These forces include operational logistics, intelligence, multipurpose canine (K9) teams, and firepower control teams, as well as communications specialists, who sustain worldwide special operations missions as directed by Commander, Marine Corps Forces Special Operations Command (COMMARFORSOC).

Marine Special Operations School

The MSOS was activated at Camp Lejeune on March 11, 2007. The mission of the school is to screen, assess, select, and train Marines and sailors for special operations assignments in MARSOC. The MSOS is the training and education proponent in support of MARSOC, providing advanced individual special operations training. The school is tasked with the mission of conducting the following programs: assessment and selection (A&S); entry-level special operations training for both noncommissioned officers (NCOs) and company grade officers for special operations asssignments in MARSOC; the component's exercise program; advanced and specialty courses; and development of MARSOF standards, doctrine, tactics, techniques, and procedures. The MSOS also serves as the liaison for training and education between MARSOC and SOCOM schools.

Marine Special Operations Intelligence Battalion (MIB)

The mission of the MIB is to train, sustain, and maintain combat readiness, and provide intelligence support at all operational levels in support of MARSOF training and operations around the world with mission-specific intelligence capability. Since its inception on August 27, 2009, the command participated in more than twenty-eight deployments to thirteen countries in support of MARSOC missions. The MARSOC Intelligence Battalion produces the direct support teams, SOTF (special operations task force) enhancements, and other capabilities as needed by the commander of U.S. Special Operations Command, and the commander of U.S. Marine Corps Forces, Special Operations Command.

The commander of the MIB, Lt. Col. Nicolas Vavich, stated: "Special operations are really not about going after any target at the first available time. Special operations are about choosing the right target at the time and place of our choosing to shape and influence the battle space in today's really complex and multidimensional environments. It is the exceptional Marines of Marine Intelligence Battalion, both east and west coast, who provide that capability to this component, and make MARSOC's contribution unique to the special operations community."

MARSOC is continuing its consistent, steady march to its authorized strength of 2,600 Marines by 2014. Currently MARSOC has just over five hundred trained CSOs, with an end strength of about eight hundred CSOs. MARSOC needs to create about three hundred more CSOs and about six hundred more Marines in total to reach its current authorized end strength.

MARSOC MISSION STATEMENT

U.S. Marine Corps Forces, Special Operations Command (MARSOC), as the U.S. Marine Corps component of U.S. Special Operations Command (USSOCOM), recruits, trains, organizes, equips, and, when directed by the commander, USSOCOM, deploys task-organized, scalable, and responsive U.S. Marine Corps Special Operations Forces (MARSOF) worldwide to accomplish the special operations (SO) mission assigned by the commander of USSOCOM and/or in support of combatant commanders and other agencies.

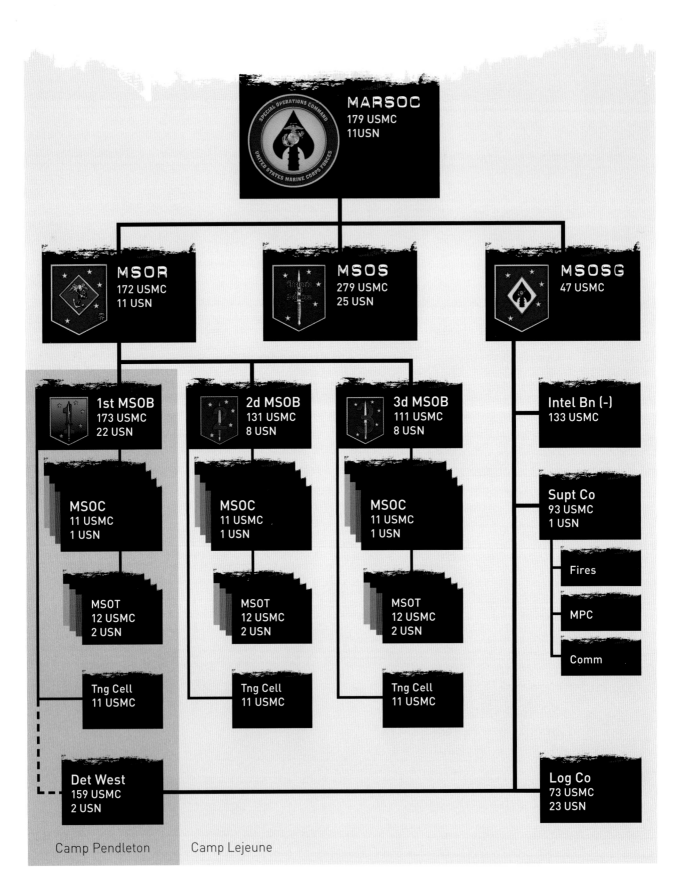

MARSOC organization (as of October 2010)

Afghan National Policemen and U.S. Special Forces engage insurgents on the far side of a canal while pulling security near the village of Hyderabad. A combined force totaling more than 350 international troops emplaced a Dry Support Bridge (DSB) across a canal near a Forward Operating Base in the area. The DSB is the first of its kind to be emplaced in Afghanistan. USMC PHOTO, SGT. BRIAN KESTER

OPERATION RED THUNDER: MARSOC IN ACTION

The village of Shewan in the Farah province of Afghanistan was a Taliban stronghold. The security situation was so bad that Gen. Stanley McChrystal, the commander, International Security Assistance Force (ISAF), could not visit the location. The Taliban had more than four hundred fighters in Shewan, and several coalition and Afghan soldiers had been killed by Taliban forces in the village. This was all about to change.

In August 2009, 1st MSOB landed in Afghanistan and joined the current MARSOC contingent operations in this area of responsibility (AOR). The new unit's arrival was in time to take part in the largest special operations to be conducted in the AOR to date. The mission was to take Shewan back from the Taliban. The mission, code-named "Operation Red Thunder," would involve 125 MARSOC/SF and more than six hundred Afghan soldiers and Afghan commandos. The Farah provincial reconstruction team, the 82nd Airborne Division, and the local Afghan National Police would provide additional support.

The mission began on September 26, 2009. During the operation, the air above Shewan was filled with AFSOC

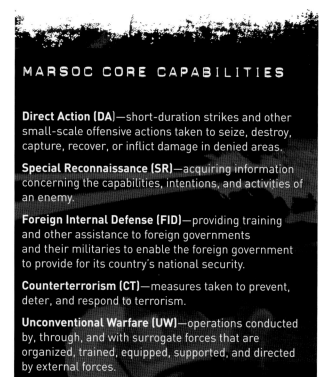

MARSOC CORE CAPABILITIES

Direct Action (DA)—short-duration strikes and other small-scale offensive actions taken to seize, destroy, capture, recover, or inflict damage in denied areas.

Special Reconnaissance (SR)—acquiring information concerning the capabilities, intentions, and activities of an enemy.

Foreign Internal Defense (FID)—providing training and other assistance to foreign governments and their militaries to enable the foreign government to provide for its country's national security.

Counterterrorism (CT)—measures taken to prevent, deter, and respond to terrorism.

Unconventional Warfare (UW)—operations conducted by, through, and with surrogate forces that are organized, trained, equipped, supported, and directed by external forces.

Information Operations (IO)—operations designed to achieve information superiority by adversely affecting enemy information and systems while protecting U.S. information and systems.

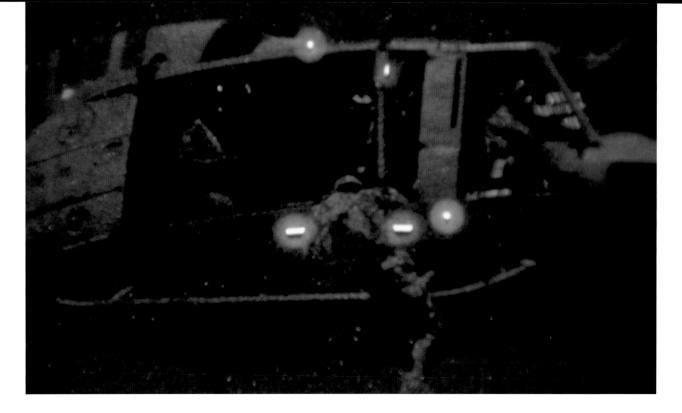

A Marine from 3rd Marine Special Operations Battalion performs a fast rope insertion from a UH-1 Huey helicopter during a maritime mobility exercise at Camp Lejeune. USMC PHOTO, SGT. EDMUND HATCH

AC-130 gunships, Army AH-64 Apaches, Marine AH-1 Cobras, Air Force A-10 Thunderbolts, and F-16 Falcon fighters providing close air support for the ground forces. The peak of the fighting occurred on September 29, so intense that the gunships overhead ran out of ammo. MARSOF and SF teams soon brought the conflict to an end through deadly and accurate small arms fire and precision air strikes. There was no doubt in anyone's mind that the air supremacy was the difference maker, and the operation was successfully completed on October 1. In four days of fighting, one Marine and three Afghan soldiers were killed in action; no civilians were killed. Sixty-five Taliban fighters were killed.

Prior to the operation in September, the MSOC commander had met with a former Taliban who warned: "In the past, many have tried to take Shewan and failed.... The Soviets tried to take Shewan and failed. The coalition forces have already tried to take Shewan and failed. You will fail if you try to take Shewan." For the men and women of SOCOM, failure is not an option, and Operation Red Thunder did not fail. The mission was such a success that the Defense Department is looking at the operation as a possible template for future operations. According to the MARSOC GFC, "The important takeaway was that we cleared the village (Afghan to USSF, 5 to 1 ratio) and held the village.... We didn't leave."

Operation Red Thunder helped clear the way for foreign and domestic aid to reach the people of Shewan and Bala Baluk district as a whole. While Shewan was still a volatile place, great progress had been made with the assistance of the Afghan National Army and the Afghan National Police, both of whom provided security for the medical outreach program.

For the first time since the Taliban had taken over, members of the Shura (village elders) were able to enter the white mosque, which had been used by the Taliban as a headquarters. On October 22, 2009, they attended a meeting there in which the Farah province governor, Rahool Amin, spoke about the rebuilding of Shewan. The governor's visit was the first of many planned visits by the local government in an effort to help Shewan remain Taliban free. As the rebuilding phase continued, there was security and safety again in Shewan, and the locals reported on the Taliban weapons caches.

The commander of the Afghan National Army conducts the preoperation briefing for Operation Red Thunder. The mission was to take the city of Shewan back from the Taliban, which had more than four hundred fighters controlling the area. MARSOC PHOTO

A MARSOC officer is the ground force commander (GFC) for Operation Red Thunder, which involved 125 MARSOC/SF shooters and more than six hundred Afghan soldiers and Afghan commandos. In the fall of 2009, coalition forces successfully cleared the strategic village of insurgents without a single civilian casualty. One of the things that set special operations operators apart from their conventional brethren is the fact that they sweat the details. MARSOC PHOTO

A Special Forces soldier kicks open a door during the battle in Shewan. The expertise of the American Special Operations Forces, accurate small arms fire, and precision close air support routed the Taliban fighters from the city. The battle raged on for four days, resulting in sixty-five Taliban killed. USASOC PHOTO

The three months following Red Thunder were spent rebuilding the village and jump starting the local economy. The process wasn't easy; the local populace had to be shown an attractive option to Taliban rule. Marines and soldiers worked hand in hand with the local Afghan leaders and awarded several contracts to rebuild the village and create jobs. Three schools were built, roads were repaired, and several agricultural projects were aimed at stopping the flow of illegal poppy farming. Getting a farmer to switch from profitable poppy to wheat proved to be challenging, but not impossible.

On January 22, 2010, Gen. Stanley McChrystal and his staff visited numerous sites throughout Regional Command West, including Shewan. Arriving at Shewan via CH-47 Chinook helicopter from Forward Operating Base Farah, the general said: "The last time I was here in Farah, I heard firsthand about how bad Shewan was getting. This time, I got to actually go there and see the recent progress for myself, thanks to the noteworthy clearing operation there. Your Marines have done some amazing work here and the people of Afghanistan will not forget it. The American people are proud of you."

CHAPTER 4

SELECTION AND TRAINING

IN A STATEMENT to the House Armed Services Committee in January 2007, Major General Hejlik related: *"MARSOC currently recruits, screens, assesses, and selects in service. MARSOC does not direct recruit; it is dependent on the USMC and USN for mature personnel well into their first five to ten years of service. MARSOC selection standards are set by conducting mission analysis on assigned special operations forces core tasks, developing mission essential task lists, determining what skills are required by Marines and sailors to complete those, and then determining what attributes a Marine or Sailor needs to have coming into MARSOC. The attributes MARSOC is looking for are those that enable the candidate to accept the greater responsibility of conducting special operations missions independent from large formations to a unique set of conditions and standards. MARSOC has identified applied intelligence, leadership (to include judgment, maturity, and cultural aptitude), and physical ability (including both determination and fitness) as selection criteria. MARSOC will not sacrifice quality for quantity in this process."*

The demands of MARSOC special operations assignments means they are looking for more than a "few good men." They are looking for the best of the best. The mere fact they are drawing from the Corps itself, where every Marine is a rifleman, gives them a leg up on most of the other services. Currently, all USMC military occupational specialties (MOSs) are eligible to apply for service with MARSOC. To fill the ranks with the right Marines starts with a thorough screening process designed to ensure that those who get assigned to MARSOC meet or exceed the established prerequisites.

The screening is a three-phase process which includes: (1) records screening, ensuring the candidate meets the MARSOC prerequisites and is medically qualified to participate in the assessment and selection (A&S); (2) physical screening consisting of physical fitness testing and swim assessment; and (3) intelligence testing, psychological evaluation, and final medical screening once the Marine reports to A&S. While all Marines seeking assignment to MARSOC undergo the screening process, only those Marines who want to train as CSOs must attend the A&S.

Critical Skills Operators (CSOs) are the Marines who conduct the real-world, no-kidding missions. They are the boots on the ground, the door kickers and shooters. They make up the MSOTs. MARSOC Marines receive specialized training in accordance with their assigned special operations core tasks. The Marines work closely with U.S. Army Special Forces and Navy Special Warfare Command SEALs to ensure MARSOC Marines and sailors have skills fully integrated and interoperable on the battlefield. Those Marines who make it as operators possess the essential combination of maturity, mental agility, physical strength, and motivation. CSO Marines have a minimum tour requirement of five years within the command.

OPPOSITE: The final exam on the Individual Training Course (ITC) is the Derna Bridge exercise. Similar to the Special Forces' Robin Sage exercise, Derna Bridge requires the Marines to utilize all of the skills taught them during the course. Upon successful completion of the ITC, the Marines become critical skills operators, and are assigned to one of the three Marine Special Operations Battalions. USMC PHOTO, LANCE CPL. THOMAS W. PROVOST

Direct Combat Support (DCS), as the name implies, provides direct support to the CSOs in a combat environment. These support functions include intelligence, infantry/recon, logistics, field artillery, explosive ordinance disposal (EOD), signals intelligence (SIGINT), and military police dog handlers. Although they are not required to attend A&S, they will often deploy with the operators into combat. For this reason, they are required to go through the same screening process as the CSO Marines.

Combat Service Support (CSS) are active duty and reserve Marines who fill billets within MARSOC. Depending upon mission requirements, they may or may not deploy with the unit. Members of the CSS are not required to attend the A&S.

Regarding whether duty with MARSOC is a career path for a Marine or a rotational assortment, Maj. Jeff Landis, MARSOC PAO, stated: *"Due to the nature of our business and the natural occurrence of rotations, it is important to maintain a strong, capable team. Team cohesion and expertise are integral to success in any special operations force and can only be built over time. With the strategic*

ABOVE: Marines who are in a direct combat support role will provide the intelligence, logistics, and a myriad of other functions for the MSOT. While they are not required to go though the assessment and selection process, they will go through the same screening process as the critical skills operators. Students in the Network Operators Course communicate with operators in the field during the course's final exercise. This photo was taken during the fourth iteration of the thirteen-week course. USMC PHOTO, LANCE CPL. KYLE MCNALLY

OPPOSITE: Critical skills operators (CSOs) are the Marines who conduct the real-world, no-kidding missions. They are the boots on the ground, the door kickers and shooters. These special operations Marines work closely and often in tandem with other SOCOM units, such as U. S. Army Special Forces and U.S. Navy SEALS.

Marines in U.S. Marine Corps Forces, Special Operations Command's new Assessment and Selection Preparation and Orientation Course conduct a ten-mile hike at Camp Lejeune, North Carolina. ASPOC is designed to prepare critical skills operator candidates for the challenges of assessment and selection. USMC PHOTO, LANCE CPL. KYLE MCNALLY

complexities of the environments and the need to maintain continuity in the teams, it is also important to have longer tour lengths than the standard three-year tour. Therefore, MARSOC personnel are not bound by traditional tour lengths. We have said for several years now, that we would like to create an MOS that provides our operators a defined career path—we think this is an important part of our continued growth in terms of capacity and the ability to attract and keep well qualified and competent Marines, offering them a clear career path, with growth, opportunities for promotion, etc. As for rotations, we are retaining our experienced and trained Marines for longer tours as a result of Headquarters, Marine Corps authorization for Marines to remain beyond the 60-month tours with incentives for Marines reenlisting and remaining in MARSOC."

Once a Marine has passed the screening process, he will then be invited to attend assessment and selection. Held five or six times a year, A&S requires the Marine to endure nineteen days of physical and mental challenges while under constant evaluation by the training cadre. The assessment enables the MARSOC instructors to identify those Marines who have the necessary attributes to complete the follow-on training and conduct special operations missions. MARSOC teams will often operate in austere environments, under adverse conditions, and in hostile territory. They must be mature, physically fit, and able to make the right decisions at a moment's notice. Candidates going through the A&S process are evaluated on a variety of criteria such as mental aptitude, swimming, land navigation, ruck marches, and so on. Each day the instructors look for the Marine who has the

One of the demanding physical challenges for Marines in ASPOC is a three hundred-meter swim. USMC PHOTO, LANCE CPL. THOMAS W. PROVOST

MARSOC CREED

My name is Marine. My title is MARSOC Silent Warrior, and I exemplify both with equal vigor and determination. My judgment, initiative, and professionalism are the hallmarks of my position.

Always Faithful, Always Forward in the fight to defend our Nation and our way of life. I will never surrender, I will never fall back, and I will always remember the Marine Warriors who have gone before me. I will defend, with my life, the honor and legacy of our great fighting spirit.

Realizing that the battlefield is always changing, I will train with fervor and intensity to ensure that I will excel in any environment and in all conditions. I will set the example for all Warrior Marines. My skill and knowledge will inspire them to follow me.

Semper Fidelis will be the guiding principle in my life and chosen profession. I will be faithful to my God, my Corps, my family, and my comrades. I will never fail those who guide me, support me, love me, and fight with me.

Outstanding leadership in combat, training, in the field, formation in garrison, and on liberty. I will lead Marines and ensure that they are ready, relevant, courteous, and respectful. By my own example and with concerned leadership, I will forge the next generation of Special Operations Forces warrior leaders.

Committed Special Operations Forces Marine tested by trial and examination. I am a Silent Warrior. I am a Marine by the Grace of our God.

Land navigation is emphasized throughout MARSOC training. The Marines are outfitted with the latest GPS equipment, but batteries fail at the most inopportune times and rucksacks may have to be dumped if escape and evasion (E+E) becomes necessary. For this reason the Marines turn to the time-honored practice of using a map and lensatic compass.

"right stuff," the operator who will be able to meet the demands of MARSOC.

The A&S process is so challenging that the Marine Special Operations School implemented a new training course. The Assessment and Selection Preparation and Orientation Course (ASPOC) serves as the precursor to the Assessment and Selection Course. This three-week course is designed to enhance the candidates' physical capabilities and prepare them mentally for A&S. With the current OPTEMP, some of the Marines are coming directly from warfighting to selection. This new course gives them the opportunity to acclimate to the new environment prior to the actual A&S process.

INDIVIDUAL TRAINING COURSE

After selection, the Marine will be assigned to the individual training course (ITC). The ITC is a seven-month course in which the candidates will be challenged anew both physically and mentally. The product of the ITC is the critical skills operator (CSO), who will be able to operate across the spectrum of MARSOC core missions, "in every clime and place." The course is set up to simulate the rigors, stresses, and complexities of combat while under the close inspection of the evaluating instructors. The ITC is broken down into four phases of training.

A Marine with the Marine Special Operations Company, 1st Marine Special Operations Battalion, assesses a simulated head wound on a role-player during a nighttime raid through a village. During the Individual Training Course (ITC), the Marines are instructed in tactical combat casualty care. USMC PHOTO, LANCE CPL. STEPHEN C. BENSON

ABOVE: In the first phase of ITC, the emphasis is on physical training, fitness, and endurance. This includes teaching the students hand-to-hand combat. Phase One concentrates on the skills and physical demands that the operators will experience during actual special operations missions. USMC PHOTO

Phase One—lasts ten weeks and is devoted to the basic skill set required by all special operations forces. The physical training (PT) program is designed around endurance, functional fitness, and amphibious training, with an emphasis placed on physical fitness, swimming, and hand-to-hand combat. This PT continues throughout the course to prepare the operator for the demands of a special operations environment. Phase One also concentrates on field skills, such as land navigation; individual and small unit tactics; patrolling; Survival, Escape, Resistance, and Evasion (SERE); and Tactical Combat Casualty Care (TCCC). Rounding out the phase will be training in mission planning, fire support, and communications.

OPPOSITE: All Marines are riflemen first and have been through the School of Infantry. In the ITC, instructors will go over small unit tactics and take the students to the next level and beyond. Operating in small fourteen-man teams, behind enemy lines, and in austere conditions, MARSOC operators have little room for error. Instructors adopt the adage, "the more one sweats in training the less you will bleed in battle." USMC PHOTO, MARSOC PAO

During demolitions training, a Marine attaches the blasting cap at the end of a timed fuse. The fuse is attached to three detonation cords connected to three separate explosives for simultaneous detonation during the practical application portion of the demolition and logistics. USMC PHOTO, LANCE CPL. STEPHEN C. BENSON

Phase Two—lasts eight weeks and concentrates on small unit tactics. Building on the foundation of Phase One, the students receive training in small boat and scout/swimmer operations, crew served weapons, demolitions, photography, and collecting and reporting of intelligence. The culmination of this phase of training is two full-mission profile exercises. The first is Raider Spirit, a two-week field exercise in which the students will reinforce the skills learned as they conduct patrolling and combat in simulated high-stress environments. The second is Operation Stingray Fury, which focuses on urban and rural reconnaissance.

A Marine takes pictures of role-players acting as terrorists during the culmination of a three-week-long special reconnaissance training module at the Military Operations in Urban Terrain facility at Camp Lejeune, North Carolina. During the exercise, Marines trained to collect information on mock enemy forces. USMC PHOTO, CPL. RICHARD BLUMENSTEIN

A Marine reviews an image he took of a role-player acting as a terrorist during the culmination of a special reconnaissance training module at Combat Town, Camp Lejeune, North Carolina. The Marine Technical Surveillance Course lasts for seventy-two days. During the exercise, Marines are trained to collect information on mock enemy forces, using digital imaging and computers.

ABOVE: From the beginning of their training, all Marines are riflemen first. Those Marines who have been selected and continue through ITC will hone their marksmanship skills. The very nature of special operations missions often relies on precise aiming and firing on the move. Here a Marine sends a few rounds downrange using his M4A1 carbine. USMC PHOTO, LANCE CPL. STEPHEN C. BENSON

RIGHT: Marksmanship does not end with the primary weapon, the carbine. There are occasions when the operator has to use his pistol. For this reason, ITC requires the students to sharpen their pistol skills. Here a Marine uses his M1911 .45-caliber semiautomatic pistol on close-in targets. USMC PHOTO, LANCE CPL. STEPHEN C. BENSON

Phase Three—lasts five weeks and covers close quarters battle (CQB), in which students train in combat marksmanship using rifles and pistols. They are taught the tactics, techniques, and procedures that they will use in assault operations as a member of a Marine special operations team. During this time the Marines also receive instruction in communications, ranging from satellite communications (SATCOM) to high-frequency radios and data systems. Phase Three ends with a series of full-mission profiles called Operation Guile Strike. These exercises will concentrate on the special reconnaissance (SR) skills that focus on precision raids on urban and rural objectives.

The Marines of MARSOC have the capability to provide the enemy with the maximum opportunity to give their life for their cause. It could be a long-range sniper round, or by calling in close air support. There are times when the work becomes up close and personal. For those occasions the Marines employ speed, surprise, and violence of action. These hallmarks of close quarter battle are emphasized during ITC. In the world of special operations missions, there are no points for second place. USMC PHOTO, MARSOC PAO

Two Marines set up a satellite communications (SATCOM) antennae during training. During ITC, students receive instruction in a wide variety of communications, ranging from SATCOM to high-frequency radios and data systems. Communications can serve as a lethal weapon or a lifeline when in battle. USMC PHOTO, CPL. KEN MELTON

Helicopter Rope Suspension Training (HRST) school teaches Marines how to insert and extract in environments where it is not possible or practical to land a helicopter. Marines fast-rope during training at Camp Lejeune, North Carolina. The training for the day is the practical application of fast rope insertion/extraction system (FRIES) from an Osprey. USMC PHOTO, LANCE CPL. DANIEL A. WULZ

Phase Four—The final phase of training presents the students with instruction on irregular warfare and lasts for seven weeks. The training ends with Operation Derna Bridge, which is similar to the Special Forces Robin Sage exercise. Derna Bridge is designed by the Marine Special Operations School to provide realistic training in irregular warfare. During this field exercise, the students will utilize all of the skills acquired throughout the ITC, requiring them to train, advise, and operate with a partner nation/irregular force. This is the final evaluation before the Marines graduate as MARSOF operators. Upon graduation, the CSO will be assigned to one of the three MSOBs.

MARSOC is steadily building its force. The goal of the command is to recruit, assess, select, and train approximately 165 CSOs each year to reach the authorized end strength of 2,600.

Advanced Training Courses

Once assigned to an MSOB, the Marines may qualify for advanced training and qualifications in combat skills, such as Airborne Training, Scuba, Advanced Language Training, CQB Leader Breacher, Advanced Sniper, Tactical Acquisition Exploitation (SR level II), Hostile Forces Tagging Tracking Locating (HFTTL), Helicopter Rope Suspension Training (HRST), Advanced EOD, Joint Terminal Attack Controller (JTAC), Unmanned Aircraft System (UAS) Operator, and specialized Survive, Evade, Resist, and Escape (SERE). A number of these advanced courses are taught at the MSOS, while others require the Marine to attend training at various bases. Some of these schools are listed as follows.

U.S. Army Airborne School, Fort Benning, Georgia. For the next three weeks, the Marines will be taught the basics needed to become "airborne" qualified. They learn the skills they will build upon to perform an insertion via parachute. Basic airborne training is broken into three weeks: Ground, Tower, and Jump Week.

Ground Week begins with an intensive program of instruction designed to prepare the Marine to complete his parachute jump. He will learn how to execute a parachute landing fall (PLF) to land safely in the LZ. Using mockups of a C-130 and a C-141, he learns the proper way to exit an aircraft.

An airborne trainee jumps from the 35-foot tower at Fort Benning, Georgia. During Tower Week, the students build confidence as they practice their parachuting skills in preparation for the actual jump. Tower Week also gives trainees practice in controlling their parachutes during the decent from the 250-foot tower and executing a parachute landing fall (PLF) upon landing. During this phase of training, the fledgling paratroopers also learn how to handle parachute malfunctions.

Airborne training for MARSOC Marines is conducted at Fort Benning. During these three weeks, the Marines learn the basic skills necessary to insert via parachute. After the successful completion of five jumps, the paratrooper earns Silver Wings. To obtain the USMC Gold Wings, they must successfully complete an additional five jumps meeting rigorous Marine Corps standards.

Next is Tower Week. Using a training device known as the swing landing tower (SLT) on which he is hooked up to a parachute harness, the trainee jumps from a twelve-foot-high elevated platform; the apparatus provides the downward motion and oscillation simulating that of an actual parachute jump. During week two the student gets to ride the Tower, which is designed to give the student practice in controlling his parachute during the decent from 250 feet, and execute a PLF upon landing.

Week three is Jump Week. The Marine will perform five parachute jumps. Upon successfully completing them, he is certified as "airborne" qualified. Upon graduation of this course, the newly frocked paratrooper wears the U.S.

Army silver jump wings, or as the Marines call them "lead wings." To obtain the Gold Wings or the Navy/Marine Parachutist Badge, he must perform an additional five jumps. These include a day and night "slick" jump, which is just with a parachute and no other equipment, and a day and night jump with full equipment. The fifth jump is usually a water jump.

The U.S. Marine Corps **Combatant Diver Course** was designed at the Naval Diving and Salvage Training Center, Panama City, Florida, in conjunction with the Marine Corps Combat Development Command, Quantico, Virginia. The course is designed by Marines, for Marines, with the purpose of providing them with

Marines may attend the U.S. Army Military Freefall Parachutist School at Fort Bragg, North Carolina. After Ground Week, students head out to Yuma Proving Grounds, Arizona. While there, each student performs a minimum of sixteen free-fall jumps that include two day and two night jumps, with oxygen and full field equipment. USMC PHOTO, LANCE CPL. STEPHEN C. BENSON

the best possible combat underwater tactical swimming training available. Emphasis is placed on developing the skills required to successfully conduct an underwater infiltration and exfiltration as required by applicable Marine Corps Orders.

Students attending the Combatant Diver Course receive training in the most current tactical doctrines and equipment, including both open- and closed-circuit diving equipment. The course is designed to provide qualified nondiving enlisted and officer personnel with the specialized training necessary to effectively operate as reconnaissance dive team members during underwater infiltration swims. To accomplish their missions, Marines must arrive undetected, on target, while keeping team integrity and maintaining the ability to execute their assigned tasks on respective shore-based objectives.

The Combatant Diver Course is thirty-five training days in length. It is divided into four modules of training: physical conditioning, combat diver fundamentals and medicine, USMC open-circuit diving equipment and operations, and USMC closed-circuit diving equipment and operations. Upon successful completion of the training, the student is certified by the Navy and Marine Corps as a USMC Combatant Diver.

The Marines also receive **Survival, Evasion, Resistance, and Escape (SERE) training** from the MARSOC Personnel Recovery (PR/SERE) branch. Those MARSOF members who will be deployed overseas are given SERE Level-C training. The nineteen-day Level-C training is an intensely physical and mentally challenging program designed to teach the students the necessary knowledge and skills to survive behind enemy lines and evade capture. Should they become prisoners, the course teaches them how to resist interrogation or exploitation, and how to escape from their captors. The Code of Conduct and survival field craft are taught with an emphasis and application to environments worldwide. Full-spectrum training includes classroom instruction, role-playing, field survival, and evasion exercises.

The **Advance Linguist Course** enhances MARSOC's ability to work bilateral missions, counterinsurgency operations, FID, and unconventional warfare (UW) operations by communications directly with host nation

Marines from U.S. Marine Corps Forces, Special Operations Command's 2nd Marine Special Operations Battalion and Marine Special Operations School step off the dock into the murky depths of Mile Hammock Bay near Camp Lejeune, North Carolina. Navy divers with 2nd MSOB and 2nd Marine Division's 2nd Reconnaissance Battalion led the training to help maintain the diver's underwater navigation skills. USMC PHOTO, LANCE CPL. STEPHEN C. BENSON

forces. All MARSOC Marines are required to undergo continual language skills training. The ALC is divided into five blocks of training, each one building upon the other. Block One begins with sixteen/twenty-four weeks of basic language skills; Block Two is a two-week iso-immersion within the continental United States to build a foundation in their assigned language; Block Three returns the students back to the classroom for twelve/eighteen weeks, expanding and improving his target language; in Block Four those students studying the CENTCOM languages go through a two-week intensive exercise, where they interact daily in role-playing scenarios focusing on those skills needed for real-world missions. In Block Five, there is review of the material and final testing.

Size up the situation, surroundings, physical condition, equipment; **U**se all your senses; **R**emember where you are; **V**anquish fear and panic; **I**mprovise and improve; **V**alue living; **A**ct like the natives; and **L**ive by your wits. Put them all together, it spells **SURVIVAL**. That is the focus of SERE training. The Survival, Escape, Resistance, and Evasion course teaches students the survival field craft skills as well as resistance, exploitation, and techniques to escape should they be captured. U.S. NAVY PHOTO, MCS1 ROGER S. DUNCAN

The **Marine Advance Sniper Course (MASC)** is a four-week course designed to train MARSOC snipers to the tactics, techniques, and procedures (TTP) in support of MARSOC missions. The MASC is comprised of five blocks of instruction: Block One covers fundamental skills, ballistics, and live fire out to two hundred meters; Block Two covers ballistics software, sniper equipment, engagement techniques, known distance shooting, and practical application of equipment; Block Three concentrates on sniper tactics, foreign weapons, shooting through glass, and shooting at moving targets at unknown distances with a variety of weapon systems from 5.56mm to .50 caliber.

Block Four is the qualification course. The student's score is based on shooting four separate and timed drills, including barricade shooting, day and night shooting, unknown distance, and spotter qualification. The last part of the training, Block Five covers a plethora of sniper TTPs: aerial sniping, explosive loophole breaching, loophole shooting, side prone, high angle, urban hides, vehicle hides, and surveillance equipment. Block Five ends with a four-day reactive exercise in which, using the full spectrum of SOF equipment and weapons, the sniper teams locate targets, conduct surveillance, and eliminate the target as part of a MSOT. Upon successful completion of the course, students are certified as MARSOC snipers.

Marine Technical Surveillance Course (MTSC) lasts for seventy-two days, providing instruction on digital collection systems, target acquisition, and exploitation through surveillance. Close-target recon applications

A Marine enrolled in the MARSOC Advanced Linguist Course takes notes in Arabic. Depending on the language, the course can last for up to a year. Dari, Pashtu, or Urdu take fifty-two weeks to complete, while French or Bahasa (the primary language of Indonesia) are thirty-six weeks in length. Not only do the Marine special operators fight in every clime and place, they are able to do so in the language indigenous to the operations area. USMC PHOTO, CPL. RICHARD BLUMENSTEIN

A MARSOC Marine sights in on his target using an M40A3 sniper rifle. The Marine Advance Sniper Course (MASC) is a four-week course designed to train MARSOC snipers in the tactics, techniques, and procedures (TTP) in support of MARSOC missions. USMC PHOTO, SGT. STEVEN KING

The MASC has five blocks of instruction that cover topics from basic fundamentals to aerial and urban environments, both day and night. Upon successful completion of the course, the operator is certified as a MARSOC sniper. Pictured here is a sniper armed with a .50-caliber M107 special application scoped rifle. USMC PHOTO, MARSOC PAO

Marines are stacked behind a Kevlar blast blanket to protect them from the explosive charge on the door. As part of the breaching course, operators learn numerous methods of entry (MOE) into various structures. Using both mechanical and explosive breaching devices, the Marines learn the best techniques to gain access to a building. USMC PHOTO, LANCE CPL. MICHAEL J. AYOTTE

provide up-to-date intelligence from boots on the ground. This bottom-up collection of intelligence and tactical analysis gives the Marines the skills to find, fix, finish, and exploit methodology to destroy terrorist/insurgent networks.

The MTSC is divided into six phases of training. During Phase One, students learn about the Advance Digital Collection Systems (ADCS). This includes data processing, computer security, digital photography, digital image editing, computer assisted drawing, Falcon View, and technical application of equipment to support all phases of an operation.

Phase Two concentrates on surveillance and counter-surveillance. Students will learn various methods of surveillance including planning, foot, vehicle, and multimode TTPs. Instruction is reinforced with practical exercises. Phase Three is close target reconnaissance (CTR), which focuses on target development, survey, and analysis. Phase Four covers method of entry (MOE), in which the students are instructed in the equipment and techniques required to obtain entry surreptitiously. Phase Five is instruction on technical application equipment, installing equipment on hosts and networks.

The course culminates with Phase Six, in which students conduct an exercise utilizing all the skills they have acquired during the course. This exercise is conducted in an area unfamiliar to the students, with a comprehensive scenario supported by role-players. During the exercise, the Marines conduct a reconnaissance patrol to a location where they spend four days operating from an observation post. There, the teams collect information to develop reports on role-players during numerous scenarios ranging from terrorist activities to normal day-to-day activities.

Members of 1st Marine Special Operations Battalion, MARSOC, are lifted from the ground by a CH-46 Sea Knight helicopter using special procedure insertion and extraction system (SPIES) rigging at Camp Margarita, aboard Camp Pendleton. The operators' training is part of MARSOC's Helicopter Rope Suspension Training Masters Course. USMC PHOTO, CPL. RICHARD BLUMENSTEIN

CHAPTER 5
WEAPONS

WHEN DET ONE (the pilot program for MARSOC) was officially activated in 2003, there were concerns that the Marines would not have the specific weapons, body armor, optics, and other personal protective equipment required for their tasks. Much of the SOCOM gear was beyond the standard USMC equipment issue and was mission essential. The same or closely similar state-of-the-art equipment would be necessary to enable any future detachment to be interoperable and sufficiently able to conduct special operations with other SOCOM units. As the detachment was standing up, it was competing with IMEF for going to war. "Gear was coming in by the truck load," Colonel Coates related. "Nobody had the authority to say 'no' to what we wanted." This chapter will cover the range of weaponry used by MARSOC.

M4A1 Carbine

Manufactured by Colt, the M4A1 is the carbine version of the full-size M16A2 assault rifle. Designed specifically for the U.S. Special Operations forces, it is likewise the primary weapon of MARSOC Marines. The M4A1's fire selector can be set for semi and fully automatic operation. The weapon is designed for speed of action and lightweight requirements, as is often the case for these CSOs. Its barrel is shortened to 14.5 inches, which reduces the weight while maintaining its effectiveness for quick-handling operations in the field. The issued collapsible buttstock has four intermediate stops, allowing adaptability in CQB without compromising shooting capabilities.

OPPOSITE: This Marine has just fired an M136 AT4 light antitank weapon during a training exercise at the Udairi Range Complex in Kuwait. With a range of 2,100 meters, the AT4's warhead is capable of penetrating 400mm of rolled homogenous armor. USMC PHOTO, SGT. BRYSON K. JONES

The M4A1 has a rifling twist rate of 1 in 7 inches, making it compatible with the full range of 5.56mm ammunitions. Its sighting system contains dual apertures, allowing for 0–200 meters, and a smaller opening for engaging targets at a longer range of 500–600 meters. Selective fire controls for the M4A1 have eliminated the three-round burst of the M16A2, replacing it with safe semiautomatic and fully automatic fire. A detachable carrying handle, when removed, exposes a Weaver-type rail for mounting a SOPMOD accessories weapon.

The M4A1 carbine is a most capable and deadly weapon suitable to the MARSOC missions. Still, further refinements were requested to make the weapon even more effective, whether for close-in engagements or against long-range

ABOVE: The primary weapon of MARSOC Marines is the M4A1 carbine. The weapon on the left is fitted with a crane stock; attached to the rail is an Elcan 1x-4x sight and AN/PVS-22 universal night sight (UNS) with 1x-12x magnification. It has a Surefire M900 vertical foregrip weapon light and an Insight visible light. The operator on the right has fitted his weapon with Magpul CTR stock; attached to the rail is a 1x-4x power scope and PEQ-15 sight. MARSOC

The M4A1 carbine is a compact version of the full-sized M16A2 rifle. The M4A1 has a fire selection for semi and fully automatic operation. The barrel has been redesigned to a shortened 14.5 inches, which reduces the weight while maintaining its effectiveness for quick-handling field operations. USMC PHOTO, MARSOC PAO

targets. To accomplish this, USSOCOM and the Naval Surface Warfare Center's Crane Division developed the SOPMOD Kit. Introduced in 1994, the special operations peculiar modification kit is issued to all U.S. Special Operations forces to expand on the capabilities and operation of the M4A1 carbine. Currently, the U.S. military is looking into further upgrading the M4A1. Whether this will result in a number of additional modifications or a brand-new weapon is still under evaluation.

M4A1 Acessory Kit

The platform for the SOPMOD kit is the rail interface system (RIS). This system replaces the front handguards on the M4A1 receiver, providing a Mil. Std. 1913 Picatinny rail. The notched rail system is located on the top, bottom, and sides of the barrel, facilitating the attaching of SOPMOD kit components on any of the four sides. The notches are numbered, making it possible to attach and reattach the various components at the same position each time it is mounted. Optical sights and night vision devices can be mounted on the top, while top and side rails would be the choice for positioning laser aiming devices or lights. The bottom of the RIS normally will accommodate the vertical grip and/or lights. When no accessories are mounted to the RIS, plastic handguards are emplaced to provide cover, and protect the unused portions of the rail. While the newer rail system is called the RAS, for rail adapter system, the RIS will accommodate a wider variety of barrels.

The M203 grenade launcher is a lightweight, single-shot, breech-loaded 40mm weapon specifically designed for placement beneath the barrel of the M4A1 carbine. With a quick-release mechanism, the addition of the M203 to the M4A1 carbine creates a versatile weapons system capable of firing both 5.56mm ammunition and an expansive range of 40mm high-explosive and special-purpose munitions. This photo also shows the crane stock.

M203 Grenade Launcher

The quick attach/detach M203 mount and leaf sight, when combined with the standard M203 grenade launcher, provides additional firepower to the operator, giving him both a point and area engagement capability. The most commonly utilized ammunition is the M406 40mm high-explosive dual-purpose (HEDP) projectile. This grenade's fragmentation effects give it a deadly radius of five meters, and it is capable of penetrating steel armor plate up to two inches thick, making it effective against both enemy personnel and light armor. Additional projectiles include M381 HE, M386 HE, M397 Airburst, M397A1 Airburst, M433 HEDP, M441 HE, M576 Buckshot, M583A1 40mm WS Para Illum, M585 White star cluster, M651 CS, M661 Green star cluster, M662 Red star cluster, M676 Yellow smoke canopy, M680 White smoke canopy, M682 Red smoke canopy, M713 Ground marker—Red, M715 Ground marker—Green, M716 Ground marker—Yellow, M781 practice; M918 target practice, M992 infrared illuminant cartridge (IRIC), 40mm nonlethal round, 40mm canister round, and 40mm sponge grenade. Future development in 40mm grenades will introduce airburst capability that will provide increased lethality and bursting radius through prefragmented, programmable HE warheads.

The quick attach M203 combines flexibility and lethality to the individual weapon. Utilizing multiple M203 setups allows concentrated fire by bursting munitions that are extremely useful in raids and ambushes. They can also illuminate or obscure the target while simultaneously delivering continuous HEDP fire. The M203 grenade leaf sight attaches to the RIS for fire control.

The M203's receiver is made of manufactured of high-strength forged aluminum alloy. This provides extreme ruggedness while keeping weight to a minimum. A complete self-cocking firing mechanism, including striker, trigger, and positive safety lever, is included in the receiver. This allows the M203 to be operated as an independent weapon, even though attached to the M16A1 and M16A2

Marines with 2nd MSOB use their rifle optics to scan the horizon during a patrol through Balu Morghab, Badghis province, Afghanistan. Patrols are conducted regularly throughout the Morghab Valley in order to actively engage the public and disrupt Taliban activity. This Marine has replaced the rail interface system with a Daniels Defense commercial off-the-shelf rail system that allows the barrel to free float, as well as giving him more real estate on which to mount accessories. CJSOTF—A, USMC PHOTO, SGT. EDMUND L. HATCH

A Marine from the 2nd MSOB returns fire on attacking Taliban fighters outside of Balu Morghab. The Taliban opened fire on a patrol consisting of 2nd MSOB Marines and Afghan National Army soldiers. This Marine is using the close quarters battle receiver (CQBR) on his M4A1. The CQBR replaces the standard barrel with a 10.3-inch barrel. This allows the Marine to work in tighter areas and on other tasks in which the compact weapon would be more practical. USMC PHOTO, SGT. EDMUND L. HATCH

rifles and M4A1 carbine. The barrel, also made of high-strength aluminum alloy, has been shortened from twelve to nine inches, allowing improved balance and handling. It slides forward in the receiver to accept a round of ammunition, then slides backward to automatically lock in the closed position, ready to fire.

Crane Stock

A modification to the M4A1 carbine is the addition of a redesigned collapsible stock designed by the Crane Division of the Naval Surface Warfare Center (NSWC) in Crane, Indiana. This stock affords the operator with a larger surface, giving him a better position for placing his cheek. This "spot weld" provides the user with a more comfortable and stabile firing position than the current tubular design. The wider stock also features storage space on both sides that can accommodate extra batteries for such items as NVGs, flashlights, GPS, etc. Worth noting are other commercial off-the-shelf (COTS) stocks like the Magpul and VLTOR, which are also popular with the special operations Leathernecks.

TARGET ACQUISITION EQUIPMENT

SU-231/PEQ

The SU-231/PEQ is part of the subprogram of the SOPMOD kit. Manufactured by EOTech, the Holographic Display Sight (HDS), as the name implies, displays holographic patterns, which have been designed for instant target acquisition under any lighting situations, without covering or obscuring the point of aim. The holographic reticle can be seen through the sight, providing the operator with a large view of the target or zone of engagement. Unlike other optics, the SU-231/PEQ is passive and gives off no telltale signature. The heads up, rectangular, full view of the HDS eliminates any blind spots, constricted vision, or tunnel vision normally associated with cylindrical sights. With both eyes open, the operator sights in on the target for true two-eyed operation.

The wide field of view of the SU-231/PEQ allows the operator to sight-in on the target/target area while maintaining peripheral viewing through the sight if needed, up to thirty-five degrees off axis. A unique feature

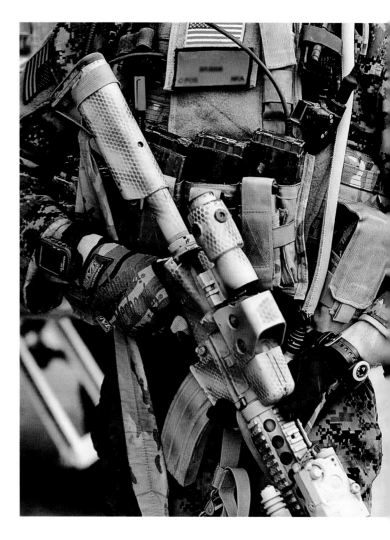

The SU-231 sight is the SOCOM-issued sight of choice for close quarters battle. The sight employs a true heads-up display that eliminates blind spots, constricted vision, and tunnel vision; true, two-eyes-open shooting is realized. Maximizing the operator's peripheral vision, and ultimately gaining greater control of the engagement zone, achieve instant threat identification. The heads-up display is constructed with a three-layer, shatterproof laminate glass that is 1/4-inch thick for added durability. Additional protection is provided by a "roll bar" ruggedized hood. MARSOC

Elcan SU-230 PVS, manufactured by Raytheon, is a switchable one power to four power dual-field-of-view (DFOV) combat optic. Using the Elcan, the operator does not have to replace optics for up close to distance shooting. With the flip of a switch the Marine can go from 1x, both eyes open, close quarters battle shooting, to 4x magnification for accurate long-range engagements, often a critical issue when operating in an urban environment.

of the HDS is the fact that it works if the heads-up display window is obstructed by mud, snow, etc. Even if the laminated window is shattered, the sight remains fully operational, with the point of aim/impact being maintained. Since many MSOT missions favor the night, it can be used in conjunction with NVG/NVD. The hallmarks of the HDS are speed and ease of use, equating incredible accuracy and instant sight-on-target operation, which can be the difference between life and death in CQB operations.

The key attribute of the HDS is extremely fast reticle-to-target acquisition in multiple target situations, and in conditions in which either the operator or the threat(s) are moving rapidly. As quickly as the eyes acquire the target, the holographic reticle can be locked onto the threat(s). When firing a weapon using the HDS, one eye maintains focus on the target while the other eye's natural instinctive reaction places the holographic reticle on the target. The result is an instant acquisition of the target for immediate and precise shot placement without covering or obscuring the point of aim. Whether engaging a target straight on, around corners or physical obstacles, or in awkward shooting positions, the HDS makes it easy for the operator to achieve rapid reticle to target lock-on. Pure and simple, the HDS locks onto the target as fast as your eyes do.

Marines take aim with their M4A1 carbines during training. The Marines' M4s are fitted with Trijicon ACOG sights. US NAVY PHOTO, MC1 TROY LATHAM

AN/PVQ-31B ACOG

The ACOG (advance combat optical gunsight), manufactured by Trijicon, is the day optical scope for the SOPMOD kit. The ACOG is a four-power telescopic sight, including a ballistic-compensating reticle. Utilizing this reticle provides increased capability to direct, identify, and hit targets to the M4A1 carbine's maximum effective range (600 meters). The sight's chevron reticle pattern provides target recognition and stand-off attack advantage while retaining a close quarters capability equivalent to the standard iron sights. The ACOG does not require any batteries.

SU-230 PVS

Another passive daylight aiming system is the Raytheon Elcan Spectre DR. Its official nomenclature is the SU-230 PVS, though the MARSOC Leathernecks simply call it the "Elcan." The SU-230 is a dual-role optical sight that incorporates both the 1x and 4x dual-field-of-view

(DFOV) in one combat optic. This switchable sight allows the operator to transition from CQB to longer range by simply throwing a lever. This will change the optic from a one power to a magnified four-power sight. The sight is designed to optimize the optical path and identical eye relief in either 1x or 4x modes. The Elcan may be used with an illuminated reticle or merely as a red dot sight. In addition to day operation, the SU-230 allows the operator to easily clip on a night vision device in front of the optic.

AN/PVS-14

The AN/PVS-14D night vision device is the optimum night vision monocular ensemble for special applications. The monocular, or pocketscope, can be handheld or placed on a face mask, helmet mounted, or attached to a weapon. The PVS-14D night vision monocular offers state-of-the-art capability in a package that meets the rigorous demands of the U.S. Special Operations forces. The monocular configuration is important to shooters who want to operate with night vision while maintaining dark adaptation in the opposite eye. The head mount assembly, a standard in the kit, facilitates hands-free operation, when helmet wear is not required. The weapon mount allows for use in a variety of applications, from using iron sights to coupling with a red dot or tritium sighting system, such as Trijicon ACOG system and EOTech HDS. A compass is available to allow the user to view the bearing in the night vision image.

Recently introduced and being evaluated is a modified PVS-14, called the AN/PSQ-20 enhanced night vision goggle (ENVG). This updated version incorporates a thermal imager into the device. The Marines will not only be able to see the enemy in the dark, but with its thermal capability, the AN/PSQ-20 ENVG will enable them to identify the heat signature of individuals and vehicles in all weather conditions, as well as degraded battlefield environments.

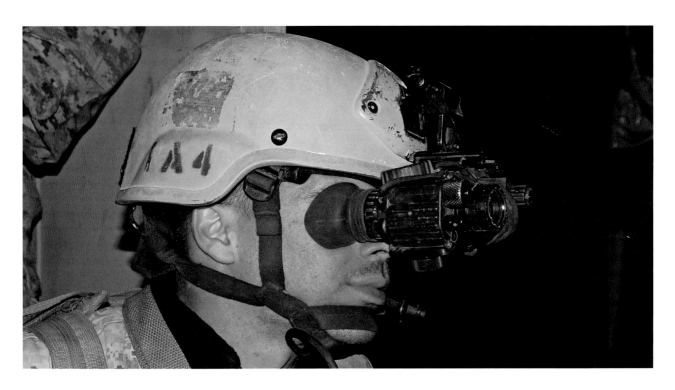

AN/PVS-14 night vision monocular may be handheld, worn on a head mount assembly, mounted on the MICH helmet, or can be mounted directly on the operator's weapon. One of the primary benefits of the PVS-14 is the fact that since it covers only one eye, the operator will have "night vision" in one eye and ambient night sight in the other. It also can be fitted on the rail system of the M4A1, providing the operator with precision aimed firing in the dark.

The AN/PVs-17 night vision sight is a replacement for the PVS-14. This sight incorporates the new Gen III image intensification I2 tube. It has a mounting attachment to interface with the Mil. Std. 1913 rail adaptor system (RAS). The system weighs less than two pounds, contains a mil-dot reticle, and can be configured for either 2.25x or 4.5x magnification.

AN/PVS-15

The AN/PVS-15 night vision device is a lightweight, self-contained Generation III twin-tube goggle system that offers the operator better depth perception than single tube systems. The PVS-15 provides similar performance characteristics as found in the AN/AVS-6 aviator's night vision imaging system (ANVIS). The AN/PVS-15 can be operated as a handheld binocular or mounted to the helmet. The AN/PVS-15 was designed specifically for times when critical mission performance and depth perception factors are required elements. The PVS-15 is powered by one AA battery and is submersible down to sixty-six feet.

AN/PVS-17

The AN/PVS-17 night vision device is a lightweight, compact, night vision sight that provides the operator the capability to locate, identify, and engage targets from twenty to three hundred meters. The MNVS features a wide field-of-view, magnified night vision image, illuminated reticle, adjustable for windage/elevation. It can be handheld or mounted on the weapon.

AN/PAS-13 (V2)

The AN/PAS-13 (V2) thermal weapons sight (TWS) is capable of detecting targets in total darkness, in adverse weather, and through other combat environment obscurities, such as dust and smoke. Placing one's eye on the sight and applying slight pressure activates the TWS, enabling the operator to detect personnel out to 1.5 kilometers and vehicles out to 4.2 kilometers. The TWS requires no visible light source to operate and can be used either as a weapon-mounted sight or handheld imager. The AN/PAS-13 IR sensor receives infrared light, which in turn converts it into digital data. It is then processed and displayed digitally as an infrared image for the operator.

The AN/PAS-13 (V2) thermal weapons sight is capable of detecting targets in total darkness, in adverse weather, and through other combat environment obscurities, such as dust and smoke.

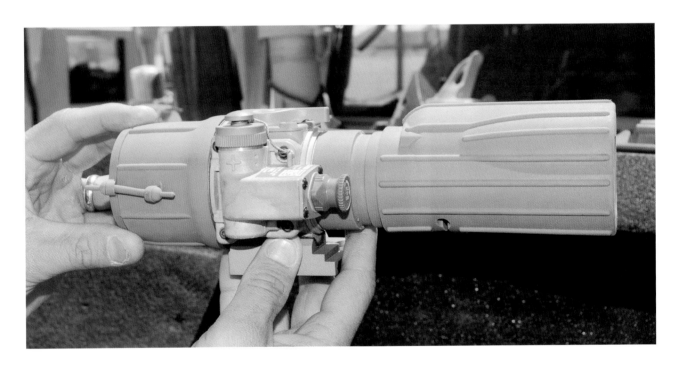

ABOVE: The AN/PVS-24, by Insight, is a clip-on night vision device that attaches to the rail system in front of the day optic. The device can quickly be mounted to the weapon without any shift in BZO (battlefield zero). This night operating device allows the Marine to observe and identify targets in adverse conditions, including rain, smoke, and snow, from low light to total darkness. Using the Gen III image tube, the device can identify targets out to six hundred meters, with the mere illumination of a quarter moon.

OPPOSITE: The Surefire M900 combines a vertical foregrip and an integral weapon light with four types of switches: a momentary-on pressure pad, a rotary constant-on switch, a system disable switch, and a momentary-on thumb switch.

AN/PEQ-15

The advanced target pointer illuminator aiming light (ITPIAL) allows the M4A1 to be effectively employed at distances of up to three hundred meters with standard-issue night vision goggles (NVG) or a weapon-mounted night vision device (i.e., AN/PVS-14). It is designed to be used with or without NVGs to engage the enemy. The IR illuminator broadens the NVG's capabilities in buildings, tunnels, jungle, overcast, and other low-light conditions in which starlight would not be sufficient to support night vision, and allows visibility in areas normally in shadow. The PEQ-15 coaligned visible and IR aiming laser combines with an adjustable and focusable IR illuminator. This combination provides the operator a decisive advantage over an opposing force with little or no night vision capability. The AN/PEQ-15, which is 50 percent smaller, replaces both the AN/PEQ-2 IR illuminator/aiming laser and the AN/PAQ-4C.

Surefire M900

The Surefire M900 is a combination vertical foregrip and integral weapon light built into one accessory. The M900 attaches to the underside of the RAS to provide the operator additional weapon control and a blazing white prefocused incandescent beam of white light. The M900 attaches to the Picatinny rail system with an ARMS M93 quick-release throw lever mount that secures the light and grip, while the M910A model attaches via a set of thumbscrews.

The weapon light features four types of switches: momentary-on pressure pads (one on each side of the grip for ambidextrous operation) that activate the main light, a rotary constant-on switch for the main light, a system disable switch to shut down the entire unit, and a momentary-on thumb switch that controls two low-output LEDs. These LEDs are available in red, white, and blue—ideal for low-signature navigation. There is an optional KT4 conversion kit that replaces the M900A's 1.62-inch

The AN/PEQ-15 advanced target pointer illuminator aiming light boosts the M4A1's range to three hundred meters with night vision devices. Smaller than previous models, it allows room for other add-ons, such as the Surefire visible flashlight shown here.

-diameter head with a 2.5-inch-diameter head that produces a narrow, farther reaching beam. The white light is bright enough to temporarily blind the night-adapted vision of an enemy combatant. The MN10 lamp assembly is rated at 125 lumens for sixty minutes.

Also available is an ultra high-output MN11 assembly rated at 225 lumens for twenty minutes. Like all Surefire lights, it is built like a tank, constructed of nonconductive Nitrolon with shock-isolated aluminum bezel. The M900A is powered by three Surefire lithium SF123A batteries that fit into the grip.

OTHER ACCESSORIES

Back-up Iron Sights (BIS)

The Back-up Iron Sight (BIS) provides the aiming ability similar to the standard iron sight on the carbine to three hundred meters. The BIS may be used in conjunction with the Aimpoint sights, and folds out of the way to allow the day optical scope or Reflex sight and night vision device to be mounted on the M4A1 carbine. In the event the optical scopes are damaged or otherwise rendered inoperable, they can be removed and the BIS will then be used to complete the mission. The sight can also be used to bore sight or confirm zero on the Reflex sight or visible laser.

Forward Handgrip

The forward or vertical handgrip, unofficially referred to as "the broom handle," attaches to the bottom of the RAS and provides added support, giving the operator a more stable firing platform. It can be used as a monopod in a supported position and allows the operator to hold the weapon despite overheating. The forward handgrip was designed for the shooter to push against the assault sling and stabilize the weapon with isometric tension during CQB; holding the weapon in this manner brings the shooter's elbows in closer or tighter to his body, consequently keeping the weapon in front of the operator, providing added support. Some of the Marines like the option; others commented they prefer using the carbine in the manner they were taught, sans the handgrip. Another preference is a modified hold that combines the more traditional handhold while pressing the heel of the supporting hand against the handgrip; this method provides a more natural stance and in turn a more stable firing platform. The forward handgrip also serves to keep the operator's hands from the handguards and barrel, which become hot during use. One of the drawbacks of the vertical grip is the possibility of the grip catching on a ledge or edge of the helicopter during egress or extraction. This issue is being addressed by the evaluation of a quick-release lever on the forward grip.

The forward handgrip is intended to mount near the center, just rear of center on the bottom rail. The forward handgrip provides added support and control for full auto and rapid firing. It allows for quicker handling when additional components are attached to the weapon. The forward handgrip can also be used as a monopod in a supporting position and allows the operator to hold the weapon despite overheating. The handgrip allows for quicker handling when the additional components have been attached to the weapon, thus providing more precise target acquisition. There are cases when the forward handgrip is necessary. One SOF master sergeant related, "Some operators end up putting so many of the [SOPMOD] accessories on the carbine, that they *have* to use the vertical grip 'cause there just isn't anywhere else to hold it!"

The Grip Pod System is an innovative vertical foregrip with a strong and stable bipod integrated into the grip. With the simple push of a button, the Grip Pod's legs instantly deploy. This unique combination combines both a vertical foregrip and bipod in a seven-ounce package. Using the Grip Pod System, the operator can go from a CQB position to prone and have a stable base for aimed firing in place before hitting the ground. It mounts to the RAS via thumbscrew and does not have a quick-release option.

Worth noting in real-world lessons learned, one of the problems encountered by soldiers in the sandbox is the pressure pad, which is most often positioned on the vertical grip (e.g., Surefire or other VFG with pressure-pad cutout). During a mission, an operator could be

A Marine with a company from 2nd MSOB fires on his target while conducting shooting drills as part of the Dynamic Assault package at the Washoe County Regional Shooting Facility, Nevada. MSOB members participated in the exercise as part of their Marine special operations companies' deployment for training (DFT). His M4A1 is fitted with a Grip-Pod forward handgrip, SU-231 sight, PEQ 15, and an Insight M3X 1913 long gun visible light. USMC PHOTO, LANCE CPL. STEPHEN C. BENSON

With the quick attach/detach sound suppressor kit MK 4 Mod 0 (QAD suppressor) in place, the report of the weapon is reduced by a minimum of 28 decibels (dB). The suppressor also reduces muzzle flash and blast substantially.

moving up an alley during a MOUT (military operations on urban terrain) operation. They are wearing NVG and moving stealthily. Without warning, the soldier trips over debris on the street, reflexively squeezes the pressure pad, and has an accidental discharge of the white light, thus giving away his position to any enemy insurgent nearby. For this reason, several of the operators have reverted back to running their white lights with the rear thumb push button.

Quick Attach Suppressor

The quick attach/detach sound suppressor can quickly be emplaced or removed from the M4A1 carbine. It measures 6.6 inches, has a diameter of 1.5 inches, and weighs less than two pounds. With the suppressor in place, the report of the weapon is reduced by a minimum of 28 decibels (dB). As the 5.56mm round is supersonic, you will hear the bang, but it is more like a .22-caliber pistol than a 5.56mm round. With the suppressor attached, it buys some time while the bad guys are trying to figure out: What was that? Where did it come from? By the time they figure out what is going on, the assault team should be in control of the situation. The suppressor will also keep the muzzle blast to a minimum, assisting the entry team in situation awareness. While the suppressor does not completely eliminate the sound, it does reduce the firing signature (i.e., the flash and muzzle blasts). Using the suppressor is effective as a deceptive measure, interfering with the enemy's ability to locate the shooter and take immediate action. Additionally, it reduces the need for hearing protection during CQB engagements, thus improving interteam voice communication.

Combat Sling

The combat sling affords a hassle-free, immediate, and secure technique of carrying the M4A1 carbine, especially when equipped with assorted accessories from the kit. The combat sling can be used alone or with the mounting hardware to provide safe and ready cross-body carry or a patrol carry. Whether moving in close quarters in a close column formation or stack, the muzzle of the weapon is kept under control and does not sweep the operator or his teammates around him. The weapon is carried in a ready position to immediately engage hostile targets. Although issued with the SOPMOD kit, various commercial manufacturers produce similar weapon slings, as well as those worked up by the company riggers, which have also found their way into the kit bags of the Marines. While there are a number of varieties of slings available, most fall into two camps single point and double point. The single-point sling provides the shooter with faster transition between shooting sides, where the two-point system is favored when the weapon must be in the carry position, such as in fast-roping or rappelling. One Marine commented, "The single point is fast, but make sure you are wearing a cup!"

Special Operations Forces Combat Assault Rifle (SCAR)

The Special Operations Forces Combat Assault Rifle (SCAR) is a modular assault rifle designed from the ground up, with input from U.S. Special Operations Forces. Designed by FN Herstal (FNH) Group headquartered in Liège, Belgium, the SCAR is manufactured at the FN Manufacturing LLC plant in Columbia, South Carolina.

The SCAR system includes the MK 16 Mod 0 (SCAR-Light) chambered in 5.56mm, MK 17 Mod 0 (SCAR-Heavy) chambered for 7.62mm, and an enhanced grenade launcher module (EGLM), which will be able to fire airburst munitions as well as standard 40mm high-explosive dual-purpose ammunition. There is a 90 percent commonality between the two SCAR versions. The SCAR-L and SCAR-H utilize all of the accessories of the SOPMOD kit (i.e., sights, lasers, scopes, etc.). To accommodate the assortment of SOPMOD accessories, the SCAR has multiple Picatinny Mil. Std. 1913 rails at three, six, nine, and twelve o'clock positions, providing the operators with multiple configurations.

The SCAR weapons employ a short-stroke, gas-piston system, eliminating the gas direct system of the M16/M4 family. This provides a more reliable weapon that does not heat up the rail system. The buttstock features a folding telescoping stock, as well as check height adjustment. This

The MK 17 SCAR-H is a selective fire, semiauto or fully auto weapon chambered for 7.62mm ammunition, in use with MARSOC in support of Operation Enduring Freedom. This Marine has fitted the weapon with an Elcan 1x-4x sight, AN/PEQ-15 ATPIAL, and Surefire M900 vertical foregrip weapon light and an Insight visible light. The collapsible stock of the SCAR can be adjusted to six positions, as well as being able to fold to the side. MARSOC

allows the operator to set the proper eye alignment to the assortment of SOPMOD optics. Both SCAR models feature three interchangeable variable-length barrels: for CQB, standard, and sniper variant. This feature was envisaged to allow the operators to fine-tune his weapon based on mission parameters.

To complement the SCAR is the MK 13 enhanced grenade launcher module (EGLM) that replaces the M203 grenade launcher. The 40mm EGLM features a left- or right-hand side opening breech to facilitate the loading of longer munitions. The EGLM includes a 40mm weapon module, Fired Control Unit I (mechanical with integrated laser/MRD sight). The launcher can be used independently or mounted onto the weapon, in similar fashion as the M203.

In August 2010 SOCOM announced it was cancelling any further acquisitions of the MK 16 version of the SCAR. However, current plans are in place to still procure the MK 17 and the MK 13 EGLM. The MK 17 is currently in use with MARSOC Marines downrange.

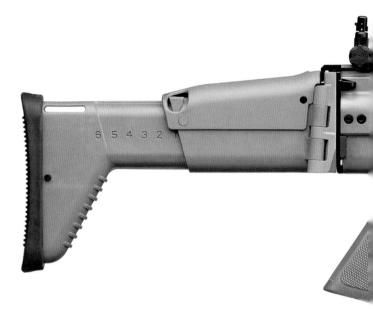

M1911

This weapon is a modified M1911A1 .45-caliber pistol, sometimes referred to as "near match" or "combat accurized." The Marines of MARSOC use what was formerly called the MEU (SOC) pistol, based on the Springfield Armory platform. With an effective range of fifty meters, this semiautomatic pistol is the designated "backup" or secondary weapon of Marines in CQB operations. Worth noting, while it is considered a secondary weapon, there are times when it would be the weapon of choice due to confined space, concealment, or other mission-dependent parameters. The M1911A1 was chosen for this role (and its modifications generated) because of its inherent reliability and lethality, and because the MEU (SOC) modifications make the M1911A1 design more "user friendly." It is reported that Col. Robert Coates, USMC commander, Det One, made the comment: "The 1911 was the design given by God to us through John M. Browning that represents the epitome of what a killing tool needs to be. It was true in 1911 and is true now."

For many decades, the Colt M1911A1 .45 ACP had been the standard handgun issued to Marines and Special Operations. With more stopping power than the 9mm Parabellum, the M1911A1 is the handgun by which all other U.S. pistols are judged. Respected for its field reliability, the single action .45's outstanding characteristics includes competition grade ambidextrous safety, precision barrel, precise trigger, rubber combat grips, Novak three-dot tritium night sights, rounded hammer spur, and an extra-wide grip safety for increased controllability. Each M1911A1 is hand-built by specialists at the Rifle Team Equipment (RTE) shop in Quantico, Virginia.

The unique characteristics of the MEU (SOC) .45 pistol are U.S. GI frame from such manufacturers as Colt, Remington Rand, Ithaca, or US&S; commercial slide from either Springfield Armory or Caspian Arms; custom-made sights by PWS; and barrel and bushing by Bar-Sto. Bar-Sto barrels have been used exclusively by the U.S. Marine Corps Marksman Unit since 1977. All barrels have a rate of twist of one turn in sixteen inches, resulting in an accuracy of two inches max at twenty-five yards using match grade ammo.

An MK 17 SCAR with the enhanced grenade launcher module single-shot, pump-action 40mm grenade launcher attached. The EGLM can be mounted beneath either of the SCAR variants (MK 16 or MK 17), or with the addition of a separate adjustable stock, the launcher can be used as a stand-alone weapon, akin to the M79. FN HERSTAL PHOTO

The M1911 is a combat accuratized modified MEU (SOC) M1911A1 .45-caliber pistol. Each pistol is hand-built by specially trained armorers at Quantico, Virginia. The base weapon is upgraded and customized until it is a match grade precision handgun. The operator carries it as a backup or secondary weapon, depending on the mission.

M1911 fitted with a Surefire X300 visible light. MARSOC is currently looking at upgrading the MEU (SOC) .45 to the M45 close quarters battle pistol. Worth noting is that the parameters for the new M45 closely mimic the M1911, including its .45 caliber; you do not fool around with perfection.

The pistol features a King's Gun Works extra-wide grip safety for increased comfort and controllability (which aids in a quick follow-up second shot) and ambidextrous thumb safety lock; it also has a Videki long aluminum trigger, Pachmayr rubber wraparound grip panels, and high-profile combat sights.

The issue magazines are replaced with Wilson Combat stainless steel, seven-round competition-grade magazines with rounded plastic follower and extended floor plate. The weapon modifications were designed in 1986 to meet the requirements of the MEU (SOC). Each pistol is hand-built by specially trained armorers at the RTE shop in Quantico, Virginia. Recent modifications in testing at the Marine Corps Systems Command, Raids and Recon in Quantico, Virginia, have been the addition of a small bracket beneath the barrel on the receiver, which allows the attaching of a visible light or laser aiming device.

M9 Beretta

Since 1985, the M9 has seen service as the standard-issue side arm for U.S. troops, both conventional and special operations forces: in Operation Urgent Fury in Grenada, Operation Desert Shield/Storm in Kuwait, Operation Restore Hope in Somalia, service with IFOR in Bosnia and KFOR in Kosovo, and now in GWOT operations. Along with the standardization of the 9mm round, the M9 brought the armed forces a larger capacity magazine. The M9 holds fifteen rounds compared to the Colt 1911's seven or eight rounds. Although the 9mm ammunition is lighter and smaller than .45 ammo, it was viewed that it was adequate for line troops. This trade-off also allowed the troops to engage more rounds in a fire fight before needing to reload. The slide is open for nearly the entire length of the barrel. This facilitates the ejection of spent shells and virtually eliminates stoppages. The open slide configuration also provides a means for the pistol to be loaded manually. Worth noting is the fact that conventional troops in general, and SOF units in particular, have voiced their dislike of the lightweight 9mm round, preferring the heavier .45-caliber ammunition. Only two units of the U.S. armed forces did not adopt the M9: Marine Force Recon companies and Delta Force.

M249SPW

The M249 special purpose weapon (SPW), which is commonly known as the Para SAW, was developed by FN Herstal to meet the requirement for a lightweight variant of the standard M249 squad automatic weapon while retaining the intrinsic functionality and reliability of the weapon. The modification resulted in a reduction in weight of 4.1 pounds, as well as a new shortened, lightweight barrel of 13.5 inches. The weapon is an individually portable, gas-operated, magazine or disintegrating metallic link-belt fed, light machine gun with fixed headspace and quick-change barrel feature. The M249SPW can engage point targets out to eight hundred meters, firing the improved NATO standard 5.56mm cartridge. The SAW forms the basis of firepower for the fire team. The gunner has the option of using thirty-round M16 magazines or linked ammunition from preloaded two-hundred-round plastic magazines. The gunner's basic

A MARSOC instructor takes aim with the M9 Beretta, firing at multiple targets during a timed shoot. Since 1985, the M9 has been the standard-issue side arm of the U.S. military. USMC PHOTO, LANCE CPL. JOSEPH STAHLMAN

The M249 Para SAW has been modified with the rail attachment system allowing the attachment of SOPMOD kit accessories, such as optical sights, night vision devices, laser designators, IR aiming devices, flashlights, and a forward pistol grip or bipod. The Para SAW is an air-cooled, belt-fed, gas-operated automatic weapon that fires from the open-bolt position. It has a cyclic rate of 750 rounds per minute (rpm). This M249 is fitted with an SU-231 sight and EOTech three-power magnifier. It is loaded with a 100-round magazine, which is often referred to as the "nut sack."

load is six hundred rounds of linked ammunition. The weapon weighs in at 19.30 pounds with a loaded two hundred-round linked ammunition magazine. Overall length is 35.75 inches, buttstock extended, and 30.50 inches with the buttstock retracted.

The Marines are evaluating the Heckler and Koch infantry automatic rifle (IAR) as a possible replacement for the M249. The M27 IAR weighs in at eight pounds when empty and eleven pounds when loaded with a thirty-round magazine. The M27 is a gas-piston system that fires from a closed-bolt position, and like the M249, it is chambered for 5.56mm NATO ammunition, making the weapon compatible with the M4A1 carbine. According to the Marine Corps System Command in Quantico, Virginia, the Corps is pursuing a high-capacity magazine that would carry fifty to one hundred rounds. Commandant General James Conway has voiced reservations regarding replacing the SAW, stating: "The M249 allows Marines to establish fire superiority in a firefight, forcing attackers to take cover. I'm going to be hard-pressed to get fire superiority over him, to keep his head down instead of him keeping mine down, because that two-hundred-round magazine just keeps on giving."

M240G Medium Machine Gun

In an effort to replace their aging stock of 7.62mm machine guns, the U.S. Marines selected the M240G medium machine gun as a replacement for the venerable Vietnam-era M60 family of machine guns. Manufactured by Fabrique Nationale, the 24.2-pound M240G medium machine gun is a gas-operated, air-cooled, link-belt fed weapon that fires the 7.62mm round. The weapon fires from an open-bolt position, with a maximum effective range of 3,725 meters. The cyclic rate of fire is 750 rounds per minute (low rate) and 950 rpm (high rate) through an adjustable gas regulator.

It features a folding bipod that attaches to the receiver, a quick-change barrel assembly, a feed cover and bolt assembly that enables closure of the cover regardless of bolt position, a plastic buttstock, and an integral optical sight rail. While it possesses many of the same characteristics as the older M60, the durability of the M240 system results in superior reliability and maintainability.

The M240G in use with MARSOC has been modified for ground use by the installation of a flash suppressor, front sight, carrying handle for the barrel, a buttstock, infantry length pistol grip, bipod, and rear sight assembly.

MK 48 Mod 1

In-theater MARSOC Marines also have access to the MK 48 Mod 1, which is a lightweight version of the M240. The FN Herstal MK 48 is essentially an M249 scaled up to accept 7.62mm ammunition. MK 48 light machine gun features a Mil. Std. 1913 rail at three, six, and nine o'clock positions, which allows for the attachment of SOPMOD accessories, such as optical sights, night vision devices, laser designators, IR aiming devices, flashlights, and a forward pistol grip or bipod. Weighing 18.26 pounds (empty), it fires the NATO standard disintegrating link-belt fed ammunition, with a cyclic rate of 750 rounds per minute.

M40A3 Sniper Rifle

In 1996, the USMC armorers at Precision Weapons Section at MCB Quantico, Virginia, began to design the replacement for the M40A1. The result was the M40A3, introduced to the Corps in 2001. It uses a Remington 700 short action, chambered for 7.62mm NATO, with a steel floorplate assembly and trigger guard built by D. D. Ross; with latter versions being replaced with Badger Ordnance hardware. The barrel is a Schneider match grade SS #7, with

OPPOSITE: A member of Marine Special Operations Company, 2nd Marine Special Operations Battalion practices high-angle fire, shooting an M240G at targets in the valley below him at the Rocket Mountain training range in Hawthorne, Nevada, in April 2009. The M240G machine gun is a 7.62mm medium class weapon manufactured by Fabrique Nationale that has been adapted from the M240/M240E1, which is a coaxial/pintle mounted machine gun for tanks and light armored vehicles. USMC PHOTO, LANCE CPL. STEPHEN C. BENSON

ABOVE: The MK 48 Mod 1 is a 7.62mm version of the MK 46, a variant of the M249SPW. The MK 48 has the versatility of the Para SAW but fires hard-hitting 7.62mm ammunition. FN HERSTAL PHOTO

a 1:12 twist and is twenty-four inches in length. The Unertl rings and bases have been replaced with Badger Ordnance steel rings. The rifles also come with a Harris bipod and an accessory rail, also built by G&G Machine. The fiberglass stock is a new McMillan Tactical A4, with adjustable cheek and length of pull.

The M40A3 is extremely accurate, designed for one-half minute of angle (MOA) with match ammunition. It is very rugged and was designed from the ground up to be a superb sniper rifle. Combined with the new M118LR ammo, it makes a system that is ranked with the best in the world. The rifle's magazine capacity is five rounds, with an effective range of one thousand yards.

M40A5 Sniper Rifle

The newest model in the M40 series of Marine Corps sniper rifles is the M40A5. It carries over many of the M40A3 specifications and maintains a two pound trigger pull. Like the M40A3 it is fitted with MilStd 1913 forward accessory rail and equipped with Badger Ordnance 34mm steel rings for mounting a scope. The updated and modified version features a threaded barrel tip for a Surefire muzzle break. This allows the shooter to attach a sound suppressor to the weapon.

M14/M14 Designated Marksman Rifle (DMR)

The M14 rifle was the American armed forces standard service rifle until the late 1960s when it was replaced by the M16A1. Although the M14 was replaced more than fifty years ago during the Vietnam War, it has enjoyed a resurgence among the U.S. military. One reason was that special operations teams needed to engage the enemy beyond the effective range of the M16 and now the M4A1 carbine. The M14 is a gas-operated, shoulder-fired weapon, firing a 7.62mm round from a twenty-round magazine. The rifle is capable of semiautomatic and fully automatic fire via a selector located on the right side of the weapon.

The M40A3 USMC sniper rifle is extremely accurate, very rugged, and was designed from the ground up by USMC armorers at Quantico, Virginia. The Marine in the background is firing an M40A3, while the shooter in the foreground is firing an M39 enhanced marksmanship rifle.

The rifle weighs in at eleven pounds with full magazine and sling. Its cyclic rate of fire is 750-rpm with an effective range of four hundred meters.

The DMR is a precision grade, 7.62mm, semiautomatic rifle. It is equipped with a simple mounting system that will accommodate a day optical sighting scope, the AN/PVS-4 Starlight Scope, and other night/low-level target engagement equipment. The DMR also features an operator attachable flash suppressor and uses both ten- and twenty-round magazines. The "basic" DMR (i.e., less sight, magazine, sling, basic-issue items, cleaning gear, suppressor, and bipod) weighs approximately eleven pounds. MARSOC operators, as well as Marine scout/sniper teams, use the DMR when they expect to need a rifle capable of delivering rapid, accurate fire against multiple targets at ranges out to eight hundred meters and with greater lethality than the M16A2 or M4A1 carbine.

M39 Enhanced Marksman Rifle

The M39 enhanced marksman rifle (EMR) is a semiautomatic, gas-operated rifle chambered for the 7.62x51mm NATO cartridge. It is a modified and accurized version of the M14 rifle modified by armorers at Quantico, Virginia, for Marine Corps use. It is based on the current designated marksman rifle (DMR), which it is now replacing.

The EMR is primarily used by Marine infantry by designated marksman to provide precision fire for units that do not rate a scout sniper. As a replacement for the DMR, the EMR fills the need for a lightweight, accurate weapon system utilizing a cartridge more powerful than the M16A4's standard 5.56x45mm NATO—the 7.62x51mm NATO. The EMR is also used by Marine scout snipers when the mission requires rapid accurate fire, and by Marine Corps explosive ordnance disposal teams.

A sniper and his spotter prepare a shot while taking the Mountain Scout Sniper Course at Marine Corps Mountain Warfare Training Center in Bridgeport, California. The M40A5 has been the official USMC sniper rifle since 2009. USMC PHOTO, LANCE CPL. SARAH ANDERSON

A MARSOC Marine armed with an M14 designated marksman rifle (DMR) patrols a village in Helmand province, Afghanistan, looking for Taliban fighters who attacked the MSOC unit and Afghan National Army. In addition to the "knock down" characteristics of the ammo used by the M14 DMR, the 7.62mm projectile has the ability to turn insurgent cover into mere concealment. USMC PHOTO, STAFF SGT. LUIS P. VALDESPINO JR.

MK 11 Mod 0

The 7.62mm MK 11 Mod 0 type rifle system is manufactured by Knight Manufacturing Company in Florida, and is a highly accurate, precision semiautomatic sniper rifle chamber capable of delivering its 7.62mm round well out to one thousand yards. With a half-inch MOA accuracy, the MK 11 has won acceptance in the SOF community as one of the finest semiautomatic sniper rifles in the world. The MK 11 is based on the original SR-25. This rifle appears to be an M16 on steroids. In fact, 60 percent of the parts are common with the M16 family. If an operator is familiar with the M16 or M4A1, his hands will naturally fall in place on the MK 11. From the pistol grip to the safety switch or magazine release, if you've handled an M16, you already know how to operate the MK 11. The result of this replication is a rifle that is quicker to assimilate, easier to maintain, and more seamless in transition than any other semiautomatic 7.62mm rifle in the world.

Similar to the M4A1, the MK 11 has two main sections: the upper and lower receiver. This allows for cleaning in the same manner that the troops have been familiar with since basic training. Another benefit of the receiver's breakdown is the fact that the rifle may be transported in a smaller package for clandestine activities. Once on target, the rifle is merely reassembled with no effect on the zero of the optics.

The MK 11 Mod 0 system includes a free-floating twenty-inch barrel, as well as a free-floating rail adapter system (RAS). The RAS is similar to the RIS on the M4A1. Another feature of the MK 11 is the ability to mount a sound suppresser. The muzzle blast becomes negligible and the only sound here is the sonic crack of the round going downrange.

M107 .50-Caliber Special Application Scoped Rifle (SASR)

The M107 is a semiautomatic, air-cooled, box magazine-fed rifle chambered for the .50-caliber, M2 Browning machine gun cartridge. Barrett Firearms Manufacturing of Murfreesboro, Tennessee, manufactures the weapon.

A MARSOC Marine fires his M39 enhanced marksman rifle (EMR) at targets in the valley below during training at the Rocket Mountain range southeast of Carson City, Nevada. USMC PHOTO, LANCE CPL. STEPHEN BENSON

This rifle operates by means of the short recoil principle. It features a twenty-nine-inch, free-floating fluted barrel, with a 1:15 right-hand twist.

The barrel is attached to a double spring yoke; this unique system provides the large weapon with an additional source of assistance as the barrel slides in and out, spreading the recoil throughout the weapon. This system, in addition to the standard buffer and spring arrangement in the lower receiver, prevents the shooter from getting slammed by heavy recoil with each shot. Any recoil energy that remains is softened by a thick rubber recoil pad.

The M107 is no small weapon, coming in at fifty-seven inches in length and weighing thirty pounds empty. A hefty box magazine holding ten rounds of M33 FMJ (full metal jacket), 661 grain, .50-caliber Browning machine gun (BMG) ammunition feeds the large rifle. The effective range of the M33 is two thousand yards. The standard .50-caliber BMG round has a muzzle velocity of 2,850 fps and will fire out to 6,800 yards. That is equivalent to sixty-eight football fields, laid end to end.

The pistol grip and fire/safe lever of the M107 is identical to those found on the M16/M4 rifles. Any shooter familiar

A Marine special operator fires his .50-caliber M107 special application scoped rifle (SASR) at targets in the valley below him during training at the Rocket Mountain firing range. With the M107, a well-trained marksman can effectively engage targets more than a mile away. USMC PHOTO, LANCE CPL. STEPHEN C. BENSON

Knight Armament Company's 7.62mm MK 11 Mod 0 evolved from the SR-25, a sniper rifle designed by the father of the AR15/M16, Eugene Stoner. The weapon is capable of shooting sub-MOA accuracy while engaging multiple targets. The MK 11 couples the accuracy of a bolt rifle with a semiautomatic rifle's capability of a quick follow-up shot. The MK 11 looks like an M16 and shares 60 percent of its parts with its "little brother." Note the PEQ-15 laser targeting device mounted on the rifle near the operator's left hand.

Mounted in the cupola of a GMV-M (Ground Mobility Vehicle, Marine) is an M2 .50-caliber Browning machine gun (BMG). Referred to as "the fifty" or "Ma Deuce," this heavy machine gun has been in service since World War II and continues in use around the world, including current combat operations in Afghanistan and Iraq. It is capable of taking out soft-skinned and lightly armored vehicles. The M2 can also wreak havoc on bunkers, buildings, and any insurgent occupants therein. It has a maximum effective range of two thousand meters, though it will reach out well past six kilometers.

with these military weapons, or even the AR-15 style of rifle, will feel right at home with the shooting configuration of this semiautomatic Barrett rifle.

The system comes packed in a watertight, airtight Pelican carrying case, allowing the weapon to be carried through or under the water. The basic M107 rifle is equipped with bipod, muzzle brake, carrying handle, metallic sights, and ten-round box magazine.

M2 .50-caliber Machine Gun

The Browning M2 .50-caliber machine gun has been serving the American soldiers since the late 1920s. Referred to as the "Ma Deuce" or simply the "fifty-cal," this heavy machine gun will definitely put "the fear of God" into any enemy. The M2 .50-caliber machine gun, heavy barrel, is a crew-served, automatic, recoil-operated, air-cooled machine gun. The M2 uses disintegrating metallic link-belt fed ammunition, and may be feed from either side by reconfiguring some of the component parts.

This gun has a back plate with spade grips, trigger, and bolt latch release. It may be mounted on most vehicles as an antipersonnel and antiaircraft weapon. The M2 is equipped with leaf-type rear sight, flash suppressor, and a spare barrel assembly. Associated components are the M63 antiaircraft mount and the M3 tripod, which is used for ground mounting. The weapon is rather large with a length of sixty-two inches and weight of eighty-four pounds. It has an effective range of two thousand meters, with a maximum range of over four miles. It has a cyclic rate of 450 rounds per minute.

Originally designed by John Browning toward the end of World War I, the M2 first saw combat in World War II, and continues as the U.S. armed forces' standard heavy machine gun around the world. It is also a valued part of the military armories of many foreign countries, including Great Britain, Canada, Australia, Germany, Japan, Brazil, Israel, India, South Africa, Spain, Indonesia, and Mexico.

MK 19-3 40mm Grenade Machine Gun

The MK 19 40mm machine gun is a self-powered, air-cooled, disintegrating-belt fed, blowback-operated fully automatic weapon. It is designed to deliver significant firepower against enemy personnel and lightly armored

The MK 19 40mm grenade machine gun is a self-powered, air-cooled, belt-fed, blowback-operated weapon designed to deliver decisive firepower against enemy personnel and lightly armored vehicles. At times, the MK 19 is used in place of the M2 BMG as the primary suppressive weapon for combat support. The weapon can engage enemy troops out to 2,200 meters with accurate and lethal fire.

vehicles. It can be used in place of the M2 machine gun, or to augment the heavy weapon.

The weapon fires a variety of 40mm grenades. The M430 HEDP 40mm grenade is capable of piercing up to two inches of armor and produces fragments to kill personnel within five meters and wound personnel within fifteen meters of the point of impact. Other system components include the MK 64 cradle mount, Mod 5; M3 tripod mount; and the AN/TVS-5 night vision sight. The complete weapon system weighs 137 pounds, with a rate of fire of 325 rounds per minute. The maximum range of the gun is 2,200 meters.

M136 AT4

The M136 AT4 is the Army's principal light antitank weapon, providing precision delivery of an 84mm high-explosive antiarmor warhead, with negligible recoil. The M136 AT4 is a man-portable, self-contained, antiarmor weapon consisting of a free-flight, fin-stabilized, rocket-type cartridge packed in an expendable, one-piece, fiberglass-wrapped tube. Unlike the M72 LAW (light antitank weapon), the AT4 launcher does not need to be extended before firing. When the warhead makes impact with the target, the nosecone crushes and the impact sensor will activate the internal fuse. Upon ignition, the piezoelectric fuse element triggers the detonator, initiating the main charge. This results in penetration where the main charge fires and sends the warhead body into a directional gas jet that is capable of penetrating over seventeen inches of armor plate. The aftereffects are "spalling," the projecting of fragments and incendiary effects, generating blinding light and obliterating the interior of the target.

CHAPTER 6
TACTICAL GEAR

MARSOC is part of SOCOM's goal to be the premier team of special warriors, tasked with helping accomplish the strategic objectives of the United States. The MARSOC motto—Always Faithful, Always Forward—builds on the Marine Corps motto, and means that our nation can rely on our Marines. That has real value for our country, for the Corps, and for SOCOM. As MARSOC takes its place in the history of the Corps, whenever the commander in chief of the United States may next make the request, "Send in the Marines," there is a good probability that MARSOC Marines are already there. In leading the Global War on Terrorism, they undoubtedly have to be thoroughly prepared and highly motivated, but they also have to be properly equipped. This chapter focuses on the gear that helps them live up to MARSOC, and the expectation to carry the fight wherever they are needed.

IMPROVED LOAD BEARING EQUIPMENT (ILBE)

During the Vietnam War, the Marine Corps used M1961 web gear that consisted of an equipment belt, suspenders, butt pack, ammo pouches, canteen cover, USMC first-aid kit, and first-aid/compass pouch. It was commonly referred to as "782" or "deuce" gear, so named after the form number that Marines signed acknowledging receipt of and responsibility for individual equipment like the web gear. As the years went by, and carrying methods as well as missions evolved, Marines placed extra magazines into BDU pockets, pouches, and vests. Today the improved load bearing equipment (ILBE) is a load carrying system that is designed to provide the Marine with a durable and lightweight method to transport individual combat clothing and equipment. The system is made up of two major components: the pack system and the assault load carrier (ALC).

Dependent on the mission, the MARSOC operator may choose the assault load carrier. The ALC provides easy access to ammunition and other items during the ever-changing mission profile of the MARSOC Marine.

OPPOSITE: The ground mobility vehicle (GMV) had its origins in Desert Storm when Special Operations Forces needed a more robust version of the Humvee. The GMV's cupola can mount various weapons, including the M2 .50-caliber machine gun and MK 19 automatic grenade launcher.

ABOVE: Two MARSOC operators model the latest in Afghanistan "Battle Rattle." Officially, it is referred to as individual load bearing equipment (ILBE). The operators have some latitude in the selection and placement of personal gear. The Marine on the left has mounted extra M1911 magazines on the front of the plate carrier, while his teammate has chosen to place M4 ammo in a similar manner. Radios, knives, and an assortment of pouches are positioned for ease of access during battle. The Marine on the left is also wearing a small waist belt which carries his individual first-aid kit (IFAK) for immediate access.

From the fast-paced CQB mission to lengthy special reconnaissance (SR) operations, configurable pouches may hold magazines for M4A1 5.56mm, MP5 9mm, or 7.62mm, dependent on the weapon of choice for the Marine. Small pockets and pouches are readily available to accommodate .45-caliber or 9mm pistol magazines, shotgun shells, first-aid field dressing, flex cuffs, strobes, chem lights, pressure dressings, and grenades (i.e. smoke, stun, CS gas, and fragmentation).

Each variant is of a modular nature, where the vest is made up of attachment points, such as the pouch attachment ladder system (PALS). The PALS is a grid system of webbing that allows for the modification of the Marines' vest with assorted holsters, magazine pouches, radio pockets, etc. Internal pockets allow the operators to stow maps or other gear. Additionally, these vests may have various-sized back pouches to accommodate such items as protective masks, helmets, demolition equipment, and other mission-essential items.

Recently, as seen in Operations Enduring Freedom and Iraqi Freedom, many of the MARSOC Marines are gravitating to the chest harness and armor plate carriers. A number of manufacturers (i.e. Blackhawk Industries, Eagle Industries, High Speed Gear, Tactical Tailor, etc.) feature such commercially off-the-shelf (COTS) harnesses. Larger pouches will accommodate two to three M4 thirty-round magazines, depending on the manufacturer, while some operators utilize what is called a shingle, with one magazine per insert. The other pouches will accommodate such items as hydration systems, radio, MRE, compass, laser pointer, or other mission-essential equipment.

The team's standard operating procedure (SOP) will determine the layout of the mission-essential gear. All team members are sure of each member's equipment capabilities; equipment is exchangeable, training continuity is achievable, and less variation in team equipment means fewer problems to consider. Another item commonly worn on the vests is the U.S. flag along with the operator's blood type, most often in an IR version.

Global Positioning System

How do MARSOC Marines navigate to exact locations through the mountains? How would a team conduct a strategic reconnaissance, or a reconnaissance and surveillance (R&S) in the dead of night, traversing sand dunes and wadis in the middle of an enemy desert or the mountainous terrain in Afghanistan? They use a device known as the global positioning system, or GPS. While these modern Leathernecks are schooled in land navigation using maps and the standard-issue lensatic compasses, it is equally important for the teams to be able to have pinpoint accuracy when conducting a DA mission through the desert, or across the frozen tundra, in enemy territory, in the middle of the night. They will need to know the position of a terrorist's hideout, a radar station, or perhaps the location of a high-value target (HVT) when reporting in to headquarters. The global positioning system is a collection of satellites that orbit the Earth twice a day. During this orbit they transmit the precise time, latitude, longitude, and altitude information. Using a GPS receiver, Special Operations Forces can ascertain their exact location anywhere on the Earth.

The Marines use the lightweight and compact AN/PSN-13 defense advanced GPS receiver (DAGR), which is a replacement for the AN/PSN-11 precision lightweight GPS receiver (PLGR). The DAGR is the latest version of the U.S. Department of Defense handheld GPS units. It addresses the demanding requirements of the U.S. Special Operations forces. The DAGR is a lightweight, handheld, self-contained, dual-frequency (L1/L2), SAASM-based (SAASM, for selective availability/antispoofing module, used by military global positioning system receivers to allow decryption of precision GPS coordinates) GPS receiver providing dual-frequency precise positioning service, with an easy to learn and use graphical user interface that provides real-time position, velocity, and time information.

The primary operational mission of the DAGR is to provide precision navigation and timing data to land-based warfighting operations and operations other than war. These include ground personnel, indirect fire weapon systems, and armored vehicles. The DAGR can also be used as a secondary or supplemental aid to aviation-based missions that involve operations in low-dynamic aircraft, such as helicopters, and as an aid to navigation in waterborne operations, such as for combat swimmers, submarines, and watercraft.

The Marines use the lightweight and compact AN/PSN-13 defense advanced GPS receiver (DAGR), which is a replacement for the AN/PSN-11 precision lightweight GPS receiver (PLGR). The primary operational mission of the DAGR is to provide precision navigation and timing data to land-based warfighting operations and operations other than war. The compact DAGR is easily stowed in the cavernous rucksack, or even in a pocket of the operator's assault vest.

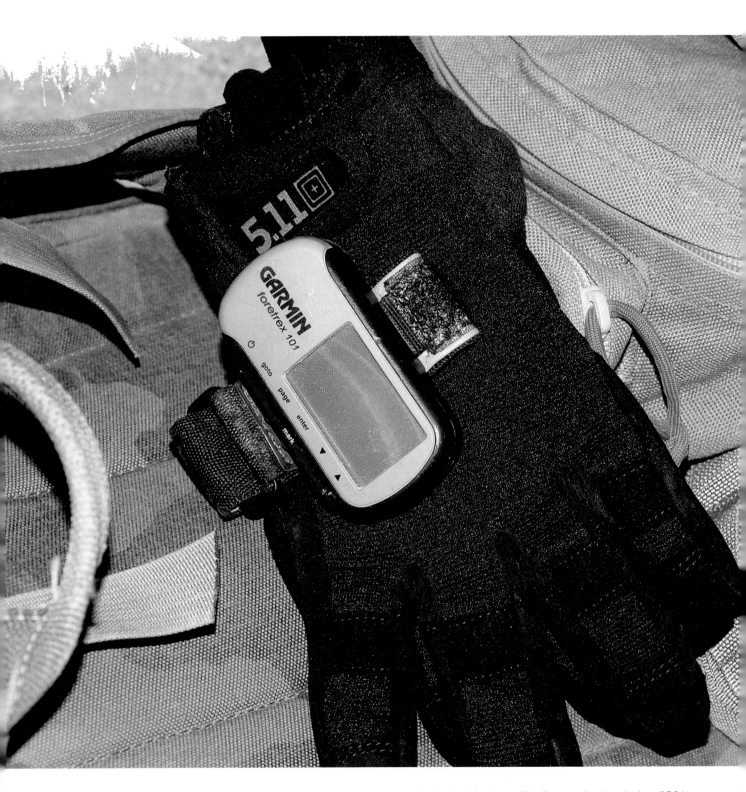

This Marine has attached a Garmin Foretrex 101 GPS navigation device to the back of his gloves. The Foretrex is a hands-free GPS in a compact, lightweight, and waterproof package. The device is waterproof, and weighing less than three ounces, it is lighter than many watches. It can store twenty routes and five hundred way points. Powered by two AAA batteries, it can run for fifteen hours.

The DAGR has new features that were not available on the older PLGR. These include web-hosted computer-based training, online status of repairs and shipping tracking, reduced reprogramming costs via web-based downloads and DAGR-to-DAGR reprogramming, enhanced antijam performance, maps functionality, graphical user interface (GUI) "ease of use" via user feedback, popular commercial GPS features and enhanced targeting, and both LRF and CAS9-line growth.

Weighing a scant one pound with batteries installed, the GPS unit is easily stowed in the cavernous rucksack or vest in use with the MSOTs. In addition to handheld operation, the DAGR can be installed in various vehicles and airborne platforms.

Strobes

Signaling has evolved immensely over the ages, from biblical times, when armies used torches, to the angle head flashlight, often referred to as the "moon beam," which has been in use since 1942. In the 1980s, chemical light sticks became popular and can still be seen in use on today's battlefield. One of the most critical needs of signaling is for the purpose of IFF, or identification, friend or foe. Installed in aircraft and combat vehicles, the IFF device emits a signal positively identifying it as friendly. For the operator on the ground, it is visible and infrared (IR) light signals.

As the MSOTs hit the battlefield today—or rather tonight—they have numerous methods of signaling, both

Infrared (IR) strobes, from left to right, MS-2000(M), Phoenix Jr. and VIPIR. Each of these strobes is omnidirectional. The MS white light is equivalent to 250,000 lumens. It is also fitted with an IR shield, visible only with NVGs. This Phoenix Jr. IR transmitter with battery weighs a mere two ounces and is user programmable. With NVGs, it is visible up to five miles. The VIPIR utilizes LEDs (light emitting diodes), and depending on the particular model, the light may have from one to five of these LED lights.

visible and infrared. The MS-2000(M) strobe is the most common among the U.S. military. Powered by two AA batteries, this strobe can be configured in white light, IR filter visible only with night vision goggles, and a blue filter with a directional shield that helps distinguish the strobe from ground fire.

The VIP and VIPIR series have been modified to include interactive friend or foe capabilities. This series of lights can be used for everything from personal identification to landing zone lighting. These signaling devices utilize LEDs (light emitting diodes), which are solid-state semiconductor diodes that emit colored light. An epoxy resin is used to encapsulate the semiconductor, while producing a lens to further focus the light.

The smallest of these is the Phoenix IR transmitter, which can still be found in use. It weighs a mere two ounces and is powered by a nine-volt battery. The Phoenix can be programmed to various lighting patterns, while the Phoenix Jr. simply has an on/off switch.

The AN/PAS-23 mini thermal monocular (MTM) is a handheld thermal imaging device. The MTM has an integrated laser pointer and digital camera. The laser pointer provides accurate target designation, while the camera enables the operator to recover, store, and download the thermal images.

When the Marines need close air support (CAS), or when they need to destroy an enemy position, they use a method called "lasing" or "painting" a target. The operator illuminates a target with a laser designator, and then the munitions (i.e., smart bomb) guide to a spot of laser energy reflected from the target. The AN/PEQ-1A SOFLAM has the capability to range out to twenty kilometers and can designate to five kilometers. USMC PHOTO, LANCE CPL. STEPHEN C. BENSON

AN/PEQ-1A SOFLAM

When it absolutely, positively has to be destroyed, you put an SOF team on the ground and a fast mover with a smart bomb in the air, and the result is one smoking bomb crater. The Special Operations Forces laser acquisition marker (SOFLAM) is lighter and smaller than the current laser marker used by regular U.S. military forces. It provides the special tactics team with the capability to locate and designate critical enemy targets for destruction using laser-guided ordnance. It can be used in daylight, or at night with the attached night vision optics.

The SOFLAM, officially called the ground laser target designator (GLTD II) by the manufacturer, Northrop Grumman, is a compact, lightweight, portable laser target designator and rangefinder. The SOFLAM is capable of exporting range data via an RS422 link and importing azimuth and elevation. It was designed to enable Special Operations Forces to direct laser-guided smart weapons, such as Paveway bombs, Hellfire missiles, and Copperhead munitions. The AN/PEQ-1A can be implemented as part of a sophisticated, digitized fire control system with thermal or image-intensified sights.

The SOFLAM uses the PRF, or pulse repetition frequency, that can be set to NATO STANAG Band I or II or is programmable. (Note: STANAG represents standards and agreements set forth by NATO, for the process, procedures, terms, and conditions under which mutual government quality assurance of defense products are to be performed by the appropriate national authority of one NATO member nation at the request of another NATO member nation or NATO organization.) PRF is the number of pulses per second transmitted by a laser.

RADIOS

Communication is paramount to a successful mission, and special operations Marines take full advantage of the latest communication gear available. You see it when they take off their floppy hats, and the headgear of the interteam radio is visible. It is there when the team unfolds the umbrella-like satellite antennae and connects it to the compact PRC-117F. Whether using UHF, VHF, line of sight, or SATCOM, the team will maintain the vital link to its headquarters.

AN/PRC-117F

Communications is the lifeline of any SOF team on a mission. For long-range communications, the AN/PRC-117F covers the entire 30 to 512 MHz frequency range while offering embedded COMSEC and Havequick I/II ECCM capabilities. This advanced software reprogrammable digital radio supports continuous operation across the 90 to 420 MHz band, providing 20 watt FM and 10 watt AM transmit power with Havequick I/II capability (10 watt FM in other frequency ranges). Fully compatible with the KY-57

The multiband AN/PRC-117F operates in VHF AM and FM, UHF AM, and UHF DAMA SATCOM. DAMA, (demand assigned multiple access) permits several hundred users to share one narrow band SATCOM channel, based on need or demand. It is voice/data and has embedded crypto, SATCOM, and ECCM capabilities. There is a GPS interface capability embedded that will transmit the user's coordinates when the operator keys the handset. The 117F provides long-range, secure communications for the Marines.

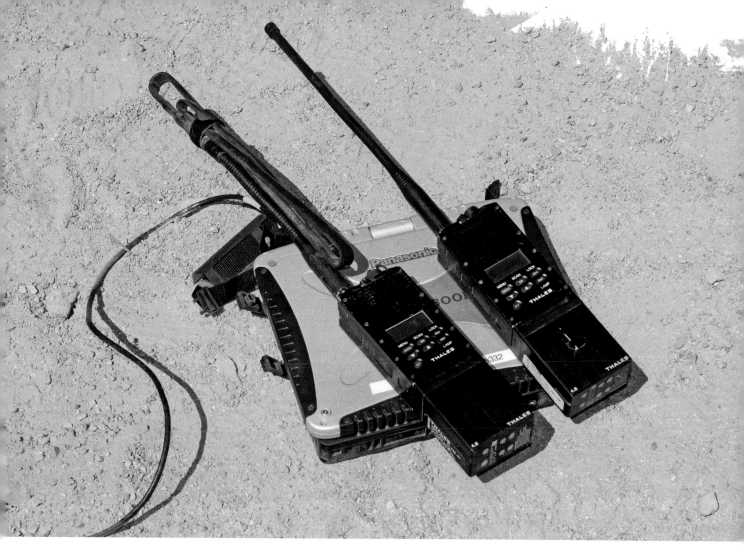

ABOVE: The AN/PRC-148 (V) maritime multiband inter/intra team radio, called the "M-Biter," is the standard team radio among U.S. SOF teams. This compact handheld individual tactical radio is AM/FM, voice or data, VHF or UHF (continuous coverage from 30-512 MHz), and is waterproof to two atmospheres. Weighing less than three pounds, it is easily stowed on the operator's assault vest.

LEFT: Intrateam communications is essential to a successful operation. The Marines wear a headset to maintain the comm link between team members. The pliable rubber ear cup establishes contact with the operator's head to lessen ambient sounds. The headset has an adjustable elastic strap, allowing the user to comfortably wear the device whether fast-roping, moving under fire, or even swimming. The unit is sealed to allow the team to insert via water.

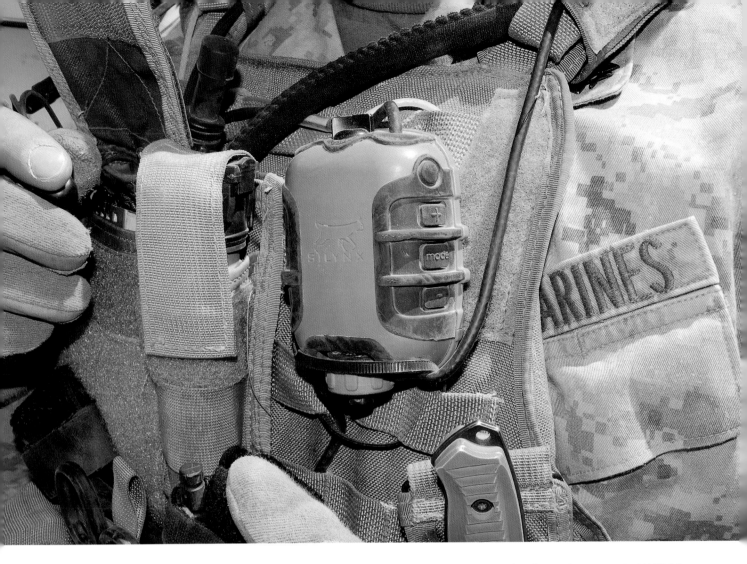

The Silynx QuietOps is an extremely versatile headset system (shown here is the push-to-talk (PTT) device). It allows the MARSOC operators to accomplish a full mission, from pre-briefing to postmission briefing, using one headset system. The Marine has the option to select an in-ear, ear muff, or covert headset, depending on the mission.

TSEC in voice and data modes to secure transmissions, the radio supports both DS-101 and DS-102 fill interfaces, and all common fill devices for Havequick word of day (WOD) and encryption key information. This device supports the Department of Defense requirement for a lightweight, secure, network-capable, multiband, multimission, antijam, voice/imagery/data communication capability in a single package.

PRC-148 MBITR

Thales Communications' multiband inter/intra team radio (MBITR) is a powerful, tactical, handheld radio designed for the U.S. Special Operations Command. The MBITR more than meets the tough SOCOM requirements, and provides a secure voice and digital-data radio with exceptional versatility, ruggedness, and reliability.

The immersible unit weighs less than two pounds and includes a keypad, graphics display, and built-in speaker-microphone. Typical of the advanced designs of Thales radios, MBITR utilizes digital-signal processing and flash memory to support functions traditionally performed by discrete hardware in other manufacturers' equipment. The power output is up to 5 watts over the 30- to 512-MHz frequency band. The MBITR has embedded Type 1 COMSEC for both voice and data traffic.

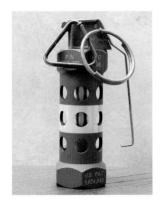

LEFT: When Marines need to make a dynamic entry into a structure, they may employ a specialized grenade known as a flashbang. The M84 is a nonfragmentation and nonlethal stun grenade that provides the entry team with a consistent and effective means of neutralizing any insurgent threat. The M84 produces a brilliant flash and deafening explosion, which disorients any terrorist in the room. This grenade gives the Marines a tactical edge when engaging enemy personnel. UNIVERSAL PROPULSION

BELOW: While the Marines of MARSOC are equipped with the latest and greatest in weaponry and equipment, the venerable smoke grenade still can be found tucked in the rucksack or chest rig. The M18 smoke grenade can be used for signaling or marking positions, such as a helicopter landing zone (HLZ), producing a thick cloud of colored smoke for fifty to ninety seconds. The grenade comes in red, green, yellow, and violet.

TACTICAL GEAR

GROUND MOBILITY VEHICLES

Thirty years ago, the U.S. Army issued a specification for a new tactical vehicle, and the concept for the high-mobility multipurpose wheeled vehicle (HMMWV or Humvee) was conceived. The new tactical vehicle entered service with the U.S. military in 1985 and quickly became as adaptable to its mission as the venerable jeep it replaced. The HMMWV has a low profile of six feet, a seven-foot-wide stance, and is fifteen feet long. This gives it a low center of gravity, making it a stable, road-hugging vehicle that is difficult to roll over. On one occasion, while traversing the edge of an impact zone, the driver told me to tighten my seat belt and headed down a hillside that I would not have tried walking down.

In the 1990s, the Special Forces missions established a need for a more robust version of the HMMWV. During the First Gulf War, the Special Forces modified Humvees for extended desert missions. These modified Desert HMMWVs were often referred to as "Dum-Vees," though this was never an official name for the vehicle. The modifications included a heavier suspension, more powerful engine, and an open bed and back for storage of water and fuel and other mission-essential items.

Further modifications of the vehicle included the placement of a cupola on the roof, which was used for mounting various weapons systems, including M2 .50-caliber machine guns, M240 7.62mm machine guns, MK 19/MK 47 40mm grenade launchers, and M249 SAW 5.56mm machine guns. Some of these modifications were conducted by the original manufacturers, while other enhancements were performed by U.S. Army depots and third party suppliers.

Used by Special Forces mounted teams, the basic deployment was four vehicles per operational detachment

The cargo compartment of the GMV-M has been modified with additional armor plating. This provides the Marines with extra protection against small arms fire and helps defeat explosively formed projectiles. Its open configuration facilitates the addition of mounts to accommodate an assortment of weapons, such as the M240 machine guns seen here. Rear doors allow the Marines to dismount quickly when necessary.

TACTICAL GEAR 135

alpha (ODA), with a crew of three soldiers per vehicle. The "Desert Humvee" greatly enhanced the capability of the mounted ODA, extending its mission endurance and flexibility. In the mid-1990s the HMMWV M1025 model was upgraded for SOF missions. These specially modified Humvees were redesignated as ground mobility vehicles, or GMVs.

The need for improved payload and performance led manufacturer AM General to provide a new expanded capacity vehicle (ECV), the model M1113. In 2003, the U.S. Special Operations Command fielded more than two hundred of the M1113 GMVs that had been modified for Special Operations Forces. In 2005, AM General began production of the M1165 variants, which incorporated higher levels of armor protection including frag kits that can be installed or removed in the field. Frag kits are vehicular armor developed by the U.S. Army Research Laboratory to defeat explosively formed projectiles (EFP), a type of improvised explosive device (IED).

Some of the modifications incorporated into the GMV include fifteen-gallon auxiliary fuel tanks to achieve longer distances, storage racks for ammo, an air compressor, electric winch, reinforcement of the rear floor, rollover bars, rear bench seats, electronic-rack mounting for communications equipment, recovery strap kits, jacks, skid plates, spare tire carriers, side rails, and an assortment of weapon mounts. Add-on armor provides 360-degree protection for the vehicle. There is also a cupola for the gunner. In addition to the weapons systems, the GMV can be fitted with smoke grenade launchers positioned in the rear and front. These tubes launch smoke canisters that airburst above and around the vehicle, creating an instant smoke screen, which is very useful when breaking contact.

Current efforts underway with the GMV family include a suspension upgrade that returns some of the payload capacity that was lost to the added armor, and a significant standardization effort to consolidate the multiple configurations of the GMV into a single SOF vehicle.

To accomplish these new upgrades, SOCOM turned to the Letterkenny Army Depot in Pennsylvania to build a vehicle with unique capabilities above and beyond those provided by the standard Army HMMWV.

A pair of GMV-Ms loaded up and ready for the next mission at a forward operating base (FOB) in Afghanistan. The Marines have added on a storage rack on the rear of the vehicle to carry rucksacks and other mission-essential gear. USMC PHOTO, MARSOC PAO

A side-by-side comparison shows the marked size difference between the HMMWV (left) and the M-ATV (right): HMMWV—sixteen feet in length, seven and a half feet wide, and six and a quarter feet in height; M-ATV—twenty and a half feet in length, eight feet wide, and nine feet in height. The mine-resistant, ambush-protected all-terrain vehicle offers more protection from roadside bombs compared to its predecessor, and features a V-shaped hull that redirects explosive blasts away from its crew. USMC PHOTO, CPL. MICHAEL CURVIN

SOCOM needed more firepower and durability than the conventional HMMWV. SOCOM wanted increased lethality, survivability, and sustainability. Working together with U.S. Special Operations Forces, the engineers and workers at Letterkenny have developed and modified special-purpose HMMWVs that are transformed into the new fighting platform called GMVs. The GMV-M (for Marine) vehicle is tailored to meet various MARSOC operational scenarios and tempos.

The U.S. Special Operations Command (SOCOM) ground mobility vehicle program modifies the basic Humvee into one of several GMV variants. The Marines of MARSOC use the GMV-M (USMC) for the Marine Special Operations Command. The workhorse GMV of the U.S. Special Operations Forces is the 1-1/2 ton, 4x4 M1165. This tactical vehicle is designed for use over all types of roads, as well as cross-country terrain, in all weather conditions. The vehicle has four driving wheels powered by a 6.5-liter V-8, turbocharged, liquid-cooled diesel engine that develops 190 horsepower at 3,400 rpm. It has four-wheel hydraulic disc brakes and a mechanical parking brake, as well as power steering. It is equipped with a pintle hook for towing; tie downs and lifting eyes are provided for air, rail, or sea shipment.

The M1165 is an expanded capacity vehicle capable of transporting a four-man crew, weapons, and mission-essential equipment. The M1165 is equipped with a reinforced frame, cross-members, lifting shackles, heavy-duty variable rate rear springs, shock absorbers, reinforced control arms, military 37x12.5 low-profile run-flat radial tires, and a transfer case and differential with a modified gear ratio to accommodate higher payloads. The suspension system is an independent coil spring-type system.

The GMV is constructed on a steel frame with boxed frame rails and five cross members constructed from high-grade alloy steel. Once the substructure is assembled, E-coating is applied to provide additional corrosion protection. The aluminum body reduces weight and

This M-ATV is equipped with a "Panama City" mine roller attachment affixed to the front of the vehicle. The device, developed by the Panama City Division of the Naval Surface Warfare Center, is designed to predetonate improvised explosive devices (IED). The vehicle has large tires, augmented hydraulics, and a durable, reinforced frame that can withstand the abuse of being driven through rugged terrain of Afghanistan. MARSOC

provides resistance to corrosion. Aluminum body panels are riveted and bonded together with technologically advanced adhesives to provide additional strength. The body is designed to flex to accommodate off-road stresses.

Fully loaded M1165s can scale grades as steep as 60 percent and negotiate side slopes of up to 40 percent. The GMV is capable of traversing hard-bottom water crossings up to thirty inches without installing a deep-water fording kit, and up to sixty inches with the kit installed. The engine is equipped for deep-water fording, having a specially sealed dipstick, dipstick tube, and vented CDR valve. Combining those features with sixteen inches of ground clearance, the GMV is an exceptional off-road vehicle.

The GMV is a light tactical vehicle specifically designed to provide a combination of mobility and payload to provide the tactical flexibility required by SOF units. The addition of field-installable armor allows the operators to trade payload for protection to meet their ever-changing mission requirements. Other upgrades include frag kits (armor protection), fire suppression systems, and a high-output 30,000 BTU air conditioning unit. The armor is available in two kits, an 'A' kit and a 'B' kit, which, when combined, provide gapless mine and ballistic protection.

The GMV is designed with an open rear, whereas the cabin of a regular HMMWV is enclosed. This flat bed area is used to store fuel, ammunition, rations, and other supplies. The GMV has a cruising range of up to 275 miles, which is well suited for Marine special operations teams that often need to operate behind enemy lines, on their own, with only occasional resupply from the air. MSOB teams, as well as other SOF units, undergo extensive training in driving and maintaining the vehicle, with an emphasis on off-road handling and in-field repairs. The new series of GMVs will serve MARSOC as they continue to prosecute America's war on terrorism.

M-ATV SOCOM VARIANT

The M-ATV is a tactical vehicle built by Oshkosh Corporation, Oshkosh, Wisconsin, that is planned to replace the up-armored HMMWV in Afghanistan. M-ATV is the mil-speak acronym for mine-resistant ambush-protected (MRAP) all-terrain vehicle. SOCOM worked closely with the MRAP Joint Program Office and Oshkosh Defense to design a vehicle that would support the unique needs of Special Operations Forces' missions. The vehicle had to provide the off-road mobility needed, as well as the lifesaving protection required. What resulted was the "tailor-made" M-ATV, with the protection of the MRAP and the mobility of the GMV.

Heavily armored MRAP vehicles, which were fielded successfully in Iraq, proved too cumbersome and unstable in the mountainous terrain of Afghanistan. The M-ATV is lighter and has the right combination of agility and armor for Afghanistan's combat and terrain demands. Despite being lighter than the MRAPs used in Iraq, the M-ATV still provides excellent protection from IEDs, roadside bombs, and rocket-propelled grenades.

The M-ATV is powered by a Caterpillar C7, 350-horsepower engine with an Allison 3500 SP transmission, giving the vehicle a top speed of sixty-five miles per hour. The vehicle incorporates rugged, durable components and systems to provide the operators the maximum in mobility and survivability. Derived from the medium tactical vehicle replacement platform (MTVR), the M-ATV incorporates an Oshkosh TAK-4 independent suspension system with sixteen inches of wheel travel that delivers the mobility needed to overcome rocky, steep terrain, mountainous landscapes, and unimproved road networks. The TAK-4 suspension uses a double-wishbone, heavy, unequal-length control-arm setup with coil springs

Marines from Special Operations Task Force West on patrol in the village of Zanghlav, Afghanistan. The coalition force conducted the patrol to increase base security and promote Government of the Islamic Republic of Afghanistan governance in Herat Province. USMC PHOTO, SGT. BRIAN KESTER

suspension with noncoaxial pneumatic shock absorbers to control vertical movement.

Noteworthy features of the M-ATV include IED jammers, V-shaped blast-dispersing monocoque hull, substantial increase in power-to-weight ratio over previous MRAPs, and the ability to ford hard-bottom freshwater to depths of up to five feet. It can generate ten kilowatts of vehicle host power and export an additional twenty kilowatts for mission equipment. The run-flat tire system enables the M-ATV to drive at thirty miles per hour, even with two flat tires. It carries up to five personnel: four plus a gunner. Its curb weight is just under twenty-five thousand pounds, with a payload capacity of four thousand pounds. The M-ATV's low center of gravity prevents rollovers and ensures maximum off-road mobility.

The SOCOM variant of the M-ATV includes a bolt-on ram bumper, which can be easily installed and removed, and a folding ladder and hood platform for easy roof access from the front of the vehicle. The cargo deck is modified to make it highly configurable in order to accept specialized equipment based on each mission's requirements. Rear storage is accessible through an armored cargo access hatch in the passenger capsule, which can seat five when configured with a common remotely operated weapon station (CROWS), or four with a gunner. The SOCOM variant also features the addition of an armored cargo hatch, providing access to mission equipment stored in the cargo deck. It will also have a larger front windshield that provides for increased visibility, particularly of threats that are above the vehicle.

The M-ATV SOCOM variant provides Marine Corps special operators the required protection combined with superior off-road mobility, which has proven beneficial during MARSOC operations in Afghanistan. DOD statistics show the casualty rate for MRAPs is 6 percent, making it "the most survivable vehicle we have in the U.S. inventory; by comparison the up-armored HMMWV has a 22 percent casualty rate."

The M-ATV is a tactical vehicle built by the Wisconsin-based Oshkosh Corporation and is planned as a replacement of the up-armored HMMWV in Afghanistan. The M-ATV is well suited to support small-unit combat operations in highly restricted rural, mountainous, and urban environments. DOD PHOTO

An M-ATV cruises down the road, its crew on the lookout for hostile targets in the windows of the buildings ahead during convoy assault training aboard Marine Corps Air Station Yuma, Arizona. USMC PHOTO, LANCE CPL. CLAYTON L. VONDERAHE

CHAPTER 7

TACTICS, TECHNIQUES, AND PROCEDURES

"TO WIN WARS BEFORE THEY EVER BEGIN" is the slogan on the MARSOC recruitment posters. As the newest member of the American special operations community, MARSOC continues to grow to its authorized strength while supporting SOCOM missions around the world. As with any new organization, there are the inevitable challenges, but nothing for which the Marines will not be able to adapt, overcome, or improvise the solutions. Commenting on the operational readiness of the Marine Special Operations Forces, Maj. Gen. Paul Lefebvre, MARSOC's third commander said: "For the first time in its history, [in 2007] MARSOC deployed a Special Operations Task Force [SOTF], into Afghanistan under CJSOTF-A. Made up of Marines, soldiers, sailors, airmen, and a variety of Department of Defense and contracted civilians, it was a critical step in the legitimacy of Marine Special Operations Forces within the SOCOM community.

"The SOTF deployment demonstrated MARSOC's ability to provide command and control of SOF over a very large battlespace, and incorporate fused intelligence capabilities, combat support and combat service support down to the lowest level." The tactics, techniques, and procedures covered in this chapter illustrate these capabilities.

ABOVE: While everyone is focused on fast-roping, MARSOC still trains in rappelling. There are times when you have to abseil down the side of a mountain, bridge, or building, and this calls for rope work, not an aerial platform. The downside of rappelling is that the Marine is connected to the rope via carabineer, which he must disconnect upon reaching the ground; the upside is that he has control of his descent and can carry a much heavier load. Pictured here, a MARSOC Marine eases down the side of a mountain during a three-day combat reconnaissance patrol in the Bala Baluk district in Afghanistan's Farah province. USAF PHOTO, STAFF SGT. NICHOLAS PILCH

OPPOSITE: The FRIES, or fast rope insertion/extraction system, is the express elevator to the battlefield. By employing the FRIES, the Force Recon team can be inserted in a matter of seconds. The FRIES begins with small woven wool cords that are braided into a rope roughly three inches in diameter. The rope is rolled into a deployment bag and the end secured to the helicopter. USMC PHOTO, MARSOC PAO

FRIES

Fast Rope Insertion/Extraction System. The way to "insert" your assault force on the ground in seconds, this system begins with small woven ropes made of wool; these ropes are then braided into a larger rope. The rope is rolled into a deployment bag and the end secured to the helicopter. Depending on the model of chopper, it can be just outside on the hoist mechanism of the side door, or attached to a bracket off the back ramp. Once over the insertion point, the rope is deployed, and even as it is hitting the ground, the Marines are jumping onto the woolen line and sliding down as easily as a fireman goes down a pole. Once the team is safely on the ground, the flight engineer or gunner, depending on the type of helicopter, will pull the safety pin, and the rope will fall to the ground. Such a system is extremely useful in the rapid deployment of MARSOC personnel; an entire assault team can be inserted within ten to fifteen seconds. FRIES is the most accepted way of getting a force onto the ground expeditiously. Unlike rappelling, once the trooper hits the ground, he is "free" of the rope and can begin his mission.

The second part of FRIES is the "extraction" method originally referred to as SPIES, or special procedure insertion, extraction system; the Army has combined both methods into one term. While fast-roping gets you down quick, there are times when you have to extract just as fast. The problem is, there is no LZ for the Blackhawk of the 160th SOAR(A) to land, and the "bad guys" are closing in on your position. This technique is similar to the McGuire and stabo rigs developed during the Vietnam War. Both used multiple ropes, which often resulted in the troops colliding into one another, though the latter at least had the benefit of allowing the user to use his weapon while on the ride up. What served the Special Forces troops of the 1960s has been refined to the new FRIES method.

While the technique has changed, the methodology remains. A single rope is lowered from the hovering helicopter. Attached to this rope are rings, woven and secured into the rope at approximately five-foot intervals. There can be as many as eight rings on the rope. The operators wearing special harnesses, similar to a parachute harness, will attach themselves to the rope via the rings.

While fast-roping gets you down quick, there are times when you have to leave just as quickly. The special patrol insertion/extraction system is used for getting "out of Dodge" in a hurry. It consists of a synthetic fiber, eight-strand braid, 1 3/4" in diameter. The tensile strength of the rope is 35,000 pounds, bolts in the sleeve are 26,000 pounds, the ring at the top of sleeve is 2,500 pounds, and the sleeve is 9,000 pounds. Here a team of Marines is attached to a line suspended beneath a V-22 Osprey. USMC PHOTO, MARSOC PAO

This is accomplished by clipping in a snap link that is at the top of the harness.

Once all team members are secured, a signal is given and the soldiers become "airborne" in reverse and extracted out of harm's way. This method, while tried and tested, allows the team members to maintain covering fire from their weapons as they extract. Once the team has been whisked out of enemy range, and an LZ can be located, the helicopter pilot will bring the troops to ground again. At this time they will disconnect from the rope and board the chopper, which will then complete the extraction.

An operator with 1st Marine Special Operations Battalion rappels out of a CH-46 Sea Knight helicopter. When inserting a team with heavy rucksack for their missions, rappelling can be the best option. Rappelling provides the operator with a smooth, stoppable transition from the aircraft to the ground. USMC PHOTO, CPL. RICHARD BLUMENSTEIN

MC-5 STATIC LINE/FREE-FALL RAM AIR PARACHUTE SYSTEM (SL/FF RAPS)

The MC-5 RAPS is the primary parachute in use as a method of insertion of Marines for deep reconnaissance or other DA missions. The MC-5 SL/FF RAPS can be configured for static line or free fall, depending on mission requirements. The MC-5 integrates components of the Paraflite MT-1XX Interim RAPS, the MC-4 RAPS, and the MT-1XS S/L convertible static line system into one versatile parachute system. The system uses identical main and reserve canopies, which not only reduce the logistics involved with separate canopies, but also eliminates the training, maintenance, and operational use of a different size/configuration reserve canopy.

The system's identical main and reserve parachutes, which can be interchanged, are 370 square feet, seven-celled, and are manufactured from 1.1-ounce F-111 nylon ripstop fabric. A cotton reinforcement/buffering panel is installed on the top and bottom skin of the canopy. The deployment sequence for the reserve canopy incorporates a free-bag system, which consists of a spring-loaded pilot chute, a bridle, and a deployment bag. The harness and container assembly has a two-pin main container to accommodate the static line. The leg straps/chest strap are fitted with quick ejector snaps.

HALO/HAHO

There are times when you cannot sneak or force your way through an enemy's front door, or even a back door or window. You must insert your team clandestinely from afar and outside of the nation's territorial airspace or boundaries. High-altitude low-opening (HALO) or high-altitude high-opening (HAHO) can be a perfect solution for MARSOC teams needing to enter hostile territory without being observed.

HALO/HAHO parachute operations start with flights over or adjacent to the designated landing area from altitudes that are much higher than those used by conventional static-line parachuting for mass infantry airdrops such as can be performed by the 82nd Airborne Division. HALO/HAHO infiltrations are normally conducted under the cover of darkness or at twilight to reduce the chance of being observed. Using the ram air parachute system (RAPS), team members deploy their parachutes at the preplanned altitude, assemble in the air, and land together at a drop zone chosen with their mission in mind. High-altitude drops can be conducted even in adverse weather conditions.

Flying at an altitude of 25,000 to 43,000 feet MSL (mean sea level) the jump aircraft, normally an MC-130 Combat Talon, appears to be a nonthreatening aircraft, perhaps a commercial airliner, on an enemy's radar screen. The radar reflection of the parachutists is too weak to register on radar. This makes military free-fall (MFF) operations ideal for the infiltration of Marine special operations teams into hostile territory. While the maximum exit altitude is 43,000 feet above sea level, MFF operations may be performed at altitudes as low as 3,500 feet above ground level (AGL).

As the aircraft approaches the insertion point, the ramp is lowered. Due to the aircraft noise combined with the MFF parachutist helmet and oxygen mask, normal verbal communication is almost impossible. This requires the team to communicate using arm-and-hand signals or interteam radios. Having already received the signals to don helmets, unfasten seat belts, and check oxygen, the jumpmaster awaits for the team to signal back "OK."

About two minutes before the aircraft reaches the insertion point, the jumpmaster raises his arm upward from his side, telling the team to "stand up." Next he extends his arm straight out at shoulder level, palm up, then bends it to touch his helmet: "move to the rear." With their rucksacks or combat assault vests loaded with mission-essential equipment, the team heads toward the rear of the plane. If jumping from the side jump door, the lead man will be a meter away from the door; if going out the rear of the plane, he will stop at the hinge of the cargo ramp. Moments turn into an eternity and then it is time. When the plane reaches the coordinates for the drop, the jump light changes to green. "Go!" In a matter of seconds, the team heads down the ramp and out into the darkness as the drone of the plane's engines fade off in

A Marine assigned to G Company, 2nd Marine Special Operations Battalion, glides through the sunset sky toward his target during military free-fall operations held at Camp Lejeune, North Carolina. The Marines are trained in high-altitude parachuting as a means of insertion into enemy territory. Exiting the aircraft while wearing oxygen and all their combat equipment, the Marines silently glide into their targets with pinpoint accuracy. USMC PHOTO, GUNNERY SGT. E. V. WALSH

the distance. Depending on the mission parameters, the team will perform a HALO or HAHO jump.

In HALO, the operators exit the plane and free fall, meeting up at a prearranged time or altitude. Jumping in this manner, the team is so small that it is virtually invisible to the naked eye, particularly under the low light of twilight or night, and of course does not show up on enemy radar screens. Using GPS units and altimeters, the team members descend until they are near the drop zone. At this point, they open their chutes and prepare for the very short remaining trip to the ground.

The HAHO alternative also starts by jumping from extreme altitude with oxygen. The difference is that as soon as the team members jump, they immediately deploy their steerable parachutes, using them to glide into the targeted area. In order to maintain formation integrity, each jumper has a strobe on his helmet, either normal or IR, and the team wears the appropriate NVGs. Additionally, each man in the team is on interteam radio for command and control of the insertion, as well as formation on the DZ.

LAR-V REBREATHER

LAR stands for lung automatic rebreather and is a closed-circuit system, meaning it does not give off any telltale bubbles to compromise the swimmer. The LAR-V MK 25 provides a combat diver with enough oxygen to stay under the water for up to four hours. The exact time depends on the individual diver's rate of breathing and his depth in the water.

The term close-circuit oxygen rebreather describes a specialized type of underwater breathing apparatus (UBA). In this type of UBA, all exhaled gas is kept within the rig. As it is exhaled, the gas is carried via the exhalation hose to an absorbent canister through a carbon dioxide–absorbent bed that removes the carbon dioxide by chemically reacting with it. After the unused oxygen passes through the canister, it travels to the breathing bag, where it is available to be inhaled again by the diver.

The gas supply used in the LAR-V is pure oxygen, which prevents inert gas buildup in the diver and allows all the gas carried by the diver to be used for metabolic needs. Closed–circuit oxygen UBA offers advantages valuable to MARSOC, including stealth infiltration, extended operating duration, and lower weight than traditional open-circuit scuba.

For underwater infiltration, Special Ops Marines use a closed-circuit oxygen underwater breathing apparatus (UBA), the Draegar LAR-V MK 25. The closed-circuit system prevents any exhaust bubbles from being seen on the surface of the water, making it easier for the team members to infiltrate to their objective without being seen by enemy forces. A MARSOC Marine lets his fellow divers know his closed-circuit oxygen rebreather is "good to go" at Mile Hammock Bay. USMC PHOTO, LANCE CPL. STEPHEN C. BENSON

HALO/HAHO techniques are a prime method of inserting a MSOT into enemy territory. Due to extreme cold encountered during high-altitude parachute operations, the jumpers must have protective clothing and equipment to shield them from the harsh environment encountered at 25,000 feet and above. USMC PHOTO, LANCE CPL. STEPHEN C. BENSON

CHAPTER 8

MARSOC INTO THE FUTURE

WHILE MARSOC CONTINUES to build up the force, it is faced with the additional challenge of concurrently deploying the force for combat and other operations around the world. Lessons learned from ongoing operations have enabled the command to adapt its structure and training, as well as the operational cycle, to improve the force's capabilities. As it brings the unit up to strength, MARSOC continues to focus on its most valuable asset: its people. Whether it is recruiting, screening, assessment, or selection of the right Marines and sailors to serve in MARSOC, the command is committed to taking care of them; they are its most critical component. By acquiring and retaining the right people with the right skill set to perform special operations, and by providing them with the very best equipment, Marine special operations teams can conduct the missions that are being assigned to them by SOCOM and the geographic combatant commanders.

The demands placed on the Marines of MARSOC are great. First and foremost, they are members of the U.S. Marine Corps, the "nation's finest." Many have pursued the path of the Amphibious Reconnaissance Marine, becoming an "elite warrior" in an already "elite force." Finally, they volunteer for MARSOC; if selected they become the crème de la crème among the Special Operations Forces of the United States military. The SOF community asks a great deal of its operators. They operate in small groups, in austere conditions, in remote areas, taking the war beyond the front lines.

ABOVE: It's an insurgent's worse nightmare as two Marines breach the door and enter the room. Most likely after blowing out the door or tossing in an M84 flashbang, the sight of these operators is the last thing an enemy fighter will ever see. As with other Special Operations Forces, MARSOC Marines plan, rehearse, and fine-tune their mission profiles. Operating at the tip of the spear, failure is not an option.

OPPOSITE: Marines and soldiers disappear into an artificial sandstorm caused by the rotor wash of a UH-60 Blackhawk helicopter. They are visiting Special Operations Task Force-West, a forward operating base (FOB) in the village of Darrah-I-Bum, Badghis Province, Afghanistan. USMC PHOTO, SGT. BRIAN KESTERAS

A pair of MARSOC Marines comes ashore with the ebb tide under the cover of fog. Operating in small teams, critical skills operators often engage an enemy force much larger than their team strength. The Marines are trained to use their brains, as well as their weaponry, to overcome these odds. USMC PHOTO, MARSOC PAO

Marines from 3rd Marine Special Operations Battalion paddle a Klepper folding canoe as part of a maritime mobility exercise at Camp Lejeune, November 18, 2010. The exercise was used by the 3rd MSOB commander to assess his battalion and quickly communicate his expectations of the command. USMC PHOTO, SGT. EDMUND HATCH

"One shot, one kill," a MARSOC Marine takes aim with a suppressed MK 11 Mod 0 sniper rifle. Every Marine is a rifleman; he stands poised to defend freedom at all costs. Those of the military know the cost of freedom more than anyone. For the MARSOC Marines, who are at the tip of the spear, they remain ever vigilant, Semper Fi! USMC PHOTO, MARSOC PAO

A sergeant with a Marine Special Operations Team using an M1911 instructs an Afghan National Army commando with an M9 Beretta on proper body positioning for firing a pistol. Outside their compound in Herat, Afghanistan, members of the 9th Commando Kandak trained alongside their Marine counterparts to sharpen their weapons skills. U.S. MARINE PHOTO, SGT. BRIAN KESTER

As members of SOCOM, the Marine special operators are not only door kickers, but are tasked with forging relationships that cross national and cultural barriers. This integration into and acceptance by local people allows MARSOC teams to have a long-term influence on the stability of a region, thus reducing the effects of religious fundamentalism and economic crises. MARSOC continues to expand its role in developing and deploying Marines and sailors with the critical skills that allow them to embrace unconventional warfare and integrate all of their capabilities to achieve success on the unconventional battlefield.

The special operations teams of MARSOC epitomize the adage, which became popular under USMC general James Mattis, "No better friend, no worse enemy." Marines who come to MARSOC are focused, purpose driven to be the best of the best, and often they are looking for a new

challenge. MARSOC is more than capable of providing all of those things, along with a new and different tool kit to meet those challenges. The specialized training they receive teaches the Marines how to use these new tools to their greatest capacity. As the United States continues to prosecute the war on terrorism around the globe, MARSOC continues to ensure that its people are prepared to accomplish the challenging and dangerous tasks to which they are assigned. MARSOC's warrior ethos and battlefield valor has earned the new unit the respect of its brothers in SOCOM. While the enemy counts how many they have killed to prove victory, MARSOC Marines focus on ensuring that their missions are accomplished with the fewest friendly casualties, and they return home.

The members of MARSOC proudly serve shoulder to shoulder with their fellow special operators of SOCOM. Yet, they remain Marines first: always faithful, always forward, silent warriors . . . oorah!

Marines with 3rd Marine Special Operations Battalion blast an opening through a mock wall during explosive breaching refresher training at Camp Lejeune. MARSOC critical skills operators constantly sharpen their skills for future operations. USMC PHOTO, LANCE CPL. THOMAS W. PROVOST

A Marine pulls security near a poppy field in Peyo as Marines and Afghan soldiers patrol through the village in the Bala Baluk district in Afghanistan's Farah province. As the sun sets, another day of fighting comes to an end. For those engaged in the global war on terrorism, it is only the day that ends. Fighting continues in the mountains of Afghanistan, deserts of Iraq, and jungles and urban battlefields around the world—everywhere terrorists call home. The Marines of MARSOC will continue to fight in "every clime and place," until the enemy is brought to justice, or justice brought to the enemy. USAF PHOTO, STAFF SGT. NICHOLAS PILCH

ACKNOWLEDGMENTS

First and foremost, I thank God, the author of liberty and freedom; may He hold our great nation in the hollow of His hand and grant our warriors victory to return home safely to their families. Thanks also to Maj. Cliff Gilmore; Maj. Michael Armistead; Maj. Jeffery Landis; Gunnery Sgt. Adrian Williams; Sgt. Edmund Hatch; Sgt. Matt Lyman; Sgt. Steven King II; Cpl. Richard Blumenstein; Cpl. Thomas Provost, MARSOC Public Affairs Office; Cynthia Hayden, MARSOC Historian, Camp Lejeune, North Carolina; Lt. Col. Tuggle; Maj. Andy Christian; Maj. Brian Fuller; Sgt. Maj. Morgan; 1st Sgt. Tim Haney, Master Sgt. Gerry, "The Punisher," 1st MSOB; Travis Haley, Magpul Dynamics; Kevin Boland, Knight's Armament Company; Lena Kaljot, U.S. Marine Corps History Division, Quantico, Virginia; Jim Ginther, Ph.D., C.A. Archives and Special Collections Branch, Library of the Marine Corps, Quantico; Barbara Sadowy, FN Herstal; Darin Anderson, Browning Advertising; Angela Harrell, Heckler & Koch; Laura Neel, Universal Propulsion; Robin Richards, Silynx Communications; Combined Joint Special Operations Task Force–Afghanistan, Public Affairs Office; International Security Assistance Forces–NATO Public Affairs Office; Capt. Kinal Sztalkoper, U.S. Army, Fort Benning; and Lisa Moore, U.S. Army Special Operations Command–Public Affairs Office, Fort Bragg.

ABBREVIATIONS

CAS	Close air support
CSAR	Combat search and rescue
CSO	Combat skills operators (a.k.a. door kickers)
CSS	Combat service support
CQB	Close quarters battle
DA	Direct action
DCS	Direct combat support
DOD	Department of Defense
DZ	Drop zone
E&E	Evasion and escape
EOD	Explosive ordnance disposal
FEBA	Forward edge battle area
FOB	Forward operation base
FOL	Forward operating location
FRIES	Fast rope insertion/extraction system
GMV	Ground mobility vehicle
GPS	Global positioning system
HAHO	High altitude high opening
HALO	High altitude low opening
HE	High explosive
HE/DP	High explosive dual purpose
HRST	Helicopter rope suspension training
HET	Human exploitation team
HLZ	Helicopter landing zone
HUMINT	Human intelligence
INTREP	Intelligence report
JCS	Joint chiefs of staff
JTAC	Joint terminal attack controller
LOS	Line of sight
LZ	Landing zone
MARSOC	Marine special operations command
MSOB	Marine special operations battalion

MSOC	Marine special operations company
MSOR	Marine special operations regiment
MSOS	Marine special operations school
MSOT	Marine special operations team
MARFORLANT	Marine forces, Atlantic
MARFORPAC	Marine forces, Pacific
MARFORRES	Marine forces, Reserve
MIB	Marine special operations intelligence battalion
MRE	Meal, ready to eat
NEC	Navy enlisted code (Navy version of MOS)
NOD	Night optical device
NVD	Night vision device
NVG	Night vision goggles
OPCON/M	Operational control/command
OPSEC	Operational security
PALS	Pouch attachment ladder system
ROC	Reconnaissance operating center
RECON	Reconnaissance
SEAL	Sea air land (U.S. Navy Special Operations Forces)
SAR	Search and rescue
SASR	Special applications scoped rifle
SERE	Survive, evade, resist, and escape
SF	Special Forces (U.S. Army)
SIGINT	Signal intelligence
SOCOM	Special Operations Command
SOF	Special Operations Forces
SOFLAM	Special Operations Forces laser acquisition marker
SPIES	Special procedure infiltration/extraction system
STX	Situational tactical exercise
SWS	Sniper weapon system
WMD	Weapons of mass destruction

TERMINOLOGY ("MARINE SPEAK")

Aft	Referring to or toward the stern (rear) of a vessel.
All hands	All members of a command.
Astern	To the rear, or behind you.
As you were	Resume activity.
Aye, Aye, Sir	Correct form to acknowledge a direct order. Naval custom.
Bivouac	Term for a camp area in the field.
Bulkhead	A wall.
C-4	Plastic explosive.
Carry on	The order to resume previous activity.
Claymore	Antipersonnel mine, 700 steel balls packed in C-4.
Cobra	Helicopter gunship.
Corpsman or Doc	A Marine or naval term for "medic."
Cover	Any form of headgear other than a helmet.
Det cord	Detonating cord; thin, flexible plastic tube packed with explosive (PETN); explodes at 25,000 fps.
Fast movers	Marine, Navy, or Air Force jet fighters or fighter/bombers.
Field Day	Thorough cleaning of a room or area.
Fighting hole	A position dug in the ground to provide cover and concealment, not the same as foxhole.
Fire team	The second smallest tactical unit in the Corps; the smallest is the individual rifleman.
FMF	Fleet Marine Force.
Galley	Shipboard kitchen; kitchen of a mess hall; mobile field kitchen.
Grunt	A Marine infantryman.
Gung-Ho	Chinese term for "working together"; understood as the team spirit.

Hatch	Door or doorway.
Hootch	Anything from a tent to a wooden hut.
Huey	Slang for UH-1 helicopter; used for troop or cargo transport.
Junk on the bunk	During inspection, a Marine lays out gear and uniforms on a flat surface, such as a bunk.
Ka-Bar	A Marine's fighting knife.
Klick	Military slang for a kilometer; 1,000 meters or 0.62 mile.
LZ	Landing zone.
MOS	Military occupational specialty; a Marine's primary training (e.g., 0300 [infantry], 0200 [intelligence], etc.).
Mustang	An enlisted Marine who has obtained an officer commission.
NCOIC	Noncommissioned officer in charge.
Oorah!	USMC response, cheer, affirmative, ready for action, etc.
Point	Also "point man"; the Marine at the head of a patrol, the first guy in line.
Rack	A bed.
Report	Presenting yourself or a group for which you are responsible to a superior.
Secure	To complete or end work on a project, or the day's work.
SitRep	Situation report.
Squared away	Someone or something that makes a good impression. Well maintained uniform or a successful exercise.
Stow	Put away.
Tail-end Charlie	Last man of a patrol; also called drag.
Topside	Upstairs.
782 gear	The equipment a Marine carries in the field, including ALICE pace, web belt, suspenders, ammunition pouches, canteens, etc. Also called deuce-gear.

INDEX

Afghanistan, 12, 43, 52, 55, 56, 59–61, 91, 92, 136, 139, 140, 151, 153, 154
aircraft
    A-10 Thunderbolt, 60
    AC-130 Spectre/Spooky gunship, 14, 15, 60
    AH-1 Cobra, 60
    AH-64 Apache helicopter, 60
    C-130, 7–8, 77
    C-141, 77
    CH-46 Sea Knight helicopter, 40, 85, 145
    CH-47, 61
    F-16 Falcon fighter, 60
    KC-130 tanker, 7, 8
    MC-130 Combat Talon, 146
    MH-47 Chinook helicopter, 12, 13
    MH-6 Little Bird helicopter, 11–13
    MH-60 Blackhawk helicopter, 12
    RH-53D Sea Stallion helicopter, 7
    UH-1 Huey helicopter, 60
    UH-60 Blackhawk, 144, 151
    V-22/MV-22/CV-22 Osprey helicopter, 8, 16, 76, 144
Amin, Rahool, 60
Antekeier, Richard P., 35
Argonaut (APS-1) submarine, 30
back-up iron sight (BIS), 102
Bainbridge, William, 23
Banana Wars, 27
Barbary Coast, Africa, 23, 24
Beckwith, Charlie, 7
Bedard, Emil R. "Buck", 43
bin Laden, Osama, 7
Borst, John L., 39
Brown, Bryan D., 43
Browning, John, 120
Camp Lejeune, NC, 18, 39, 40, 51, 56, 57, 60, 66, 72, 76, 80, 147, 152, 154
Camp Pendleton, CA, 36, 39, 44, 45, 54, 85
Carlson, Evans, 27, 30–31
Coates, Robert J., 44, 45, 47, 87, 106
Colburn, Ernest A., 35
combat sling, 104–105
Conway, James, 50, 109
Corey, Russell, 36
Decatur, Stephen, 23
Donovan, William "Wild Bill", 32
Draeger LAR-V rebreather, 51
Eaton, William, 24
Eddy, William, 32
Edson, Merritt "Red Mike", 27, 29–30, 32
Flood, James H. A., 35
Floyd, Bill, 40
Fort Benning, GA, 77, 78
Fort Bragg, NC, 10, 11, 79
FRIES, 144–145
GPS, 68, 124–127, 147
Ground Mobility Vehicle (GMV), 56, 120, 123, 134, 136–138
Guadalcanal, 29–30
HALO/HAHO, 146–148, 149
Harrington, Samuel M., 27
Hejlik, Dennis J., 49, 63

Hillard, Ron, 40
Holcomb, Thomas, 27, 31
Holland, Charles R., 43
Holloway, James L. III, 8
Howard, Marcellus, 32
Humvee, 123, 136–137, 140
Hurlburt Field, FL, 16, 47
improved load-bearing equipment (ILBE), 123–124
Jefferson, Thomas, 23
Jones, James L, 35, 43
Kaneohe Bay, HA, 41
Kelley, Paul X., 18
Knox, William Franklin, 32
Korea, 33, 35
Lakehurst Naval Air Station, NJ, 32
Landis, Jeff, 50, 64
LAR-V rebreather, 148
Lefebvre, Paul, 143
Linsday, James J., 9
M1165 vehicle, 137
MacDill AFB, FL, 9
Makin Island, 30–31
Manor, Leroy, 8
Mark V craft, 14
Mattis, James, 153
M-ATV SOCOM variant, 138–141
MC-5 RAPS, 146
McChrystal, Stanley, 59, 61
Meyers, Bruce F., 36
Miller, C. J., 27
Nautilus (SS-168) submarine, 30, 31
night vision goggles/device (NVG/NVD), 93, 95–98, 101, 104, 127, 148
O'Bannon, Presley, 24, 41
operations
    Desert Shield/Storm, 108, 123
    Eagle Claw, 7–9
    Enduring Freedom, 48, 56
    Iraqi Freedom, 46–48
    Red Thunder, 59–61
    Restore Hope, 108
    Ski Jump, 34
    Torch, 32
    Urgent Fury, 108
Ortiz, Peter, 32
Patrick, Charles, 36
Pearl Harbor, 27
Quantico, VA, 32, 79, 106–109, 111, 113
radios, 130–132
Reagan, Ronald, 9
Roosevelt, Franklin D., 27
Rumsfeld, Donald, 48
September 11 attacks, 43
Shepard, Charles Jr., 32
Shewan, Afghanistan, 59–61
Siciliano, Anthony, 44
sights, 93–102, 115
Smith, Holland, 35–36
SOC-R boats, 13
Stoner, Eugene, 120
strobes, 127–138

suppressor (for weapons), 104
target designators, 129
Taylor, Joe, 39
Tehran, Iran, 7
Tinian Island, 36
Turner, Kelly, 35–36
USS Kitty Hawk, 43
USS Maine, 24
USS Nimitz, 7
USS Perch, 34
USS Philadelphia, 23, 24
Vance, Robert, 32
Vaught, James, 9
Vavich, Nicolas, 57
Vietnam, 39–40
weapons
    105mm howitzer, 14
    25mm chain gun, 14
    75mm pack howitzer, 31
    9mm Parabellum, 106
    Bofors cannon, 14
    Browning automatic rifle (BAR), 25–27
    Crane stock, 93
    Ka-Bar knife, 36
    M1 Garand rifle, 25, 27, 31
    M107 .50-caliber special application scoped rifle (SASR), 54, 81, 116, 118, 120
    M136 AT4 light antitank weapon (LAW), 87, 121
    M14/M14 designated marksman rifle (DMR), 40, 113, 115, 116
    M16 assault rifle, 40, 87–90
    M-1903 Springfield rifle, 27
    M1911 .45-caliber pistol, 31, 44, 74, 106–108, 153
    M2 .50-caliber Browning machine gun (BMG), 120, 123, 136
    M203 grenade launcher, 89, 90
    M240G medium machine gun, 111, 136
    M249SPW (Para SAW), 108–109, 136
    M27 infantry automatic rifle (IAR), 109
    M39 enhanced marksman rifle (EMR), 113, 117
    M3A1 "grease gun", 36, 40
    M40A3 sniper rifle, 81, 111, 113
    M40A5 sniper rifle, 113
    M4A1 carbine, 41, 45, 46, 51, 74, 87–89, 92, 95, 98, 102, 103
        accessories, 102–105
    M60 machine gun, 111
    M72 light antitank weapon (LAW), 121
    M9 Beretta, 108, 153
    MK 11 Mod 0 type rifle system, 116, 119, 153
    MK 18 carbine, 10
    MK 19 40mm automatic grenade launcher, 56, 120–121
    MK 48 Mod 1 machine gun, 111
    Reising machine gun, 31
    SCAR rifle, 12, 105–107
    smoke grenade, 133, 151
    Thompson machine gun, 27, 31
Woodward, David D., 39